SUNQUAKES

PUBLISHING FOR THE WORLD
125 Years
THE JOHNS HOPKINS UNIVERSITY PRESS

SUNQUAKES

PROBING THE INTERIOR OF THE SUN

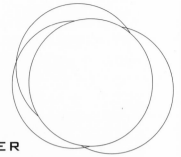

J. B. ZIRKER

THE JOHNS HOPKINS UNIVERSITY PRESS

BALTIMORE AND LONDON

© 2003 The Johns Hopkins University Press
All rights reserved. Published 2003
Printed in the United States of America on acid-free paper
2 4 6 8 9 7 5 3 1

The Johns Hopkins University Press
2715 North Charles Street
Baltimore, Maryland 21218-4363
www.press.jhu.edu

Library of Congress Cataloging-in-Publication Data

Zirker, Jack B.
Sunquakes : probing the interior of the sun / J. B. Zirker.
p. cm.
Includes index.
ISBN 0-8018-7419-X (hardcover : acid-free paper)
1. Helioseismology.   2. Convection (Astrophysics)   I. Title.
QB539.I5Z57 2004
523.7'6—dc21   2003007923

A catalog record for this book is available from the British Library.

CONTENTS

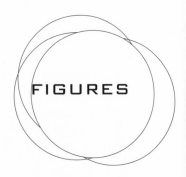

FIGURES

Figures with asterisks also appear in a color gallery following page 146

ACKNOWLEDGMENTS

ONCE AGAIN I wish to thank my editor, Trevor Lipscombe, for his encouragement and for his continuous review of the work in progress. His comments helped to make the book more accessible to the general reader.

I thank Edward Rhodes for his meticulous reading of the first draft and his valuable suggestions. John Leibacher and Frank Hill provided essential information on the genesis of the GONG (Global Oscillation Network Group) project. And to all the other researchers I mention in the book, I offer my heartfelt thanks for their contributions to this exciting science.

SUNQUAKES

THE DISCOVERY

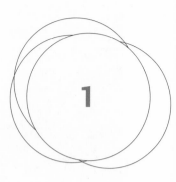

1

IN 1960, two talented scientists were working independently on similar projects. One of them would make the most important discovery in solar physics within half a century, the other would be credited with a near miss. They would learn of their blind race only when they met at a conference in Italy that year.

Both men were studying the motions of gas near the surface of the Sun. The well-known and respected solar physicist used a slow, sure, classical technique. The experimental physicist, a relative newcomer to solar physics, invented a novel way of looking at the Sun. That distinction would make all the difference to their success.

Ever since 1904, when French astronomer Pierre Janssen published his remarkable photographs of the Sun, astronomers have known that the solar surface is covered with a changing pattern of bright cells, the "granules" (fig. 1.1). Theorists such as Ludwig Biermann in Germany and Martin Schwarzschild at Princeton University explained that granules are convective cells, hot buoyant bubbles of gas that carry heat from the solar interior to the surface by rising, cooling, and sinking, like the bubbles in a pot of boiling oil.

Granules are typically about 1500 km in diameter, roughly the size of Alaska. But in angular size as seen from Earth, they are tiny, a mere arc-second or two across. That's about the size of a dime seen at a distance of 2 km. In 1960, most

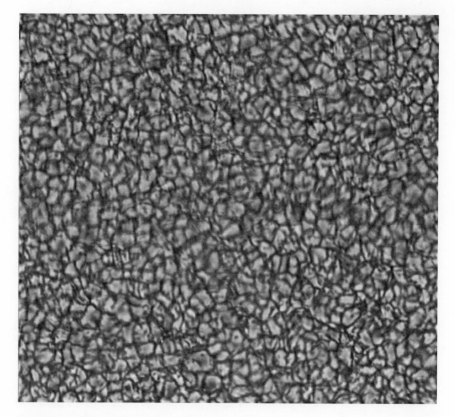

FIG. 1.1   Granules cover the visible surface of the Sun. A typical granule is about 1500 km across.

telescopes could not distinguish details on the Sun much smaller than that, because turbulence in the Earth's atmosphere blurred their images. During the day, as the atmosphere heats up, bubbles of warm and cool air mix above the telescope. Each bubble can act as a weak lens, bending light from the sun in a different direction. The net result at the telescope's focus is a blurry image. As a result, we had only fragmentary information about the brightness and motions of the granules, but because granules play a key role in transporting solar energy to space, astronomers were keen to learn more about them.

Around 1958, Schwarzschild conceived a method to avoid the troublesome turbulence. He proposed to have a balloon carry a 30 cm telescope to the stratosphere. At an altitude of 26 km, the air is so thin that its turbulence is no longer a problem. Not a balloonist himself, he and his team devised a remotely controlled telescope to take photographs of granules from the stratosphere. He launched his balloon four times and, despite some technical glitches, obtained photographs that revealed the true brightness and size of granules. However, Schwarzschild's "Project Stratoscope" left open many questions, such as the lifetimes and internal motions of granules. John Evans and Robert Leighton decided independently to study them further.

EVANS AND HIS METHOD

In 1960, Evans was the director of the Sacramento Peak Observatory, a solar observatory in the mountains of southern New Mexico. He had studied astronomy at Harvard College Observatory under the guidance of its charismatic director, Harlow Shapley. Upon receiving his doctorate in 1938, Evans taught astronomy for a year at Mills College in Oakland, California. Then, with the outbreak of World War II, he joined a team at the University of Rochester to design a variety of optical instruments for the military. This experience was crucial because it showed him that he had a gift for invention.

In the late 1940s, Donald Menzel, a Harvard professor and world-class solar physicist, convinced the United States Air Force to establish two new solar observatories, with the goal of predicting solar flares. Flares are violent explosions on the Sun that emit deadly bursts of X rays and charged electrical particles. When these emissions reach Earth, they can black out essential radio communications. The air force was therefore interested in being able to predict them.

Flares were known to originate in the low solar corona, and to study them, one had to observe the corona as often as possible. The instrument of choice was the coronagraph, a special telescope invented in 1939 by French astronomer Bernard Lyot (note 1.1). A coronagraph produces an artificial total eclipse of the Sun on demand but requires a site where the sky is deep blue right up to the Sun's edge. Such sites are normally on remote mountaintops, high up in the Earth's atmosphere and far away from the pollution of big cities.

At the time, Evans was building instruments for Walter Orr Roberts at the

High Altitude Observatory in Boulder, Colorado. Roberts, a former student of Menzel, had set up a small coronagraph at Climax, Colorado, at an altitude over 3300 meters. While Climax offered pure blue skies, it experienced fierce winter storms and heavy snowfall. Sleeping and thinking at that altitude were also difficult for visiting astronomers. Menzel and Roberts planned to replace its coronagraph with a much larger one, and to find a site for an identical instrument at a somewhat more comfortable altitude, where a permanent scientific staff could live. Evans was chosen to design and supervise the construction of both telescopes, which would be the largest in the world.

After an extensive search, Menzel and Roberts decided on Sacramento Peak as a site for the second observatory because of its deep blue skies. As an added bonus, the solar images at the site were unusually sharp because of the smooth flow of air over the peak.

Once "Sac Peak" was formally established as an observatory, a director (or "superintendent," as the air force preferred to say) was needed. Jack Evans was the obvious choice because of his broad background in optics and solar physics. Under his leadership the observatory grew rapidly in size and importance. His duties as director of a major observatory left little time for his own research. However, he pioneered in the study of the dynamics of the corona, using special narrow-band optical filters of his own design.

In 1952, he led an expedition to Khartoum in the Sudan to observe a total eclipse of the Sun. The decision to go there was no reflection on his new coronagraph, but from previous experience he knew that the sky near the Sun was a thousand times darker during a natural total eclipse than it was during an artificial eclipse, even at the best sites. During a natural eclipse, therefore, one could photograph much fainter details of the inner corona. Indeed, he returned from the Sudan with a unique set of coronal spectra and a broken arm, the latter incurred by falling off a ladder.

One of Evans's most powerful instruments was a large spectrograph that was linked to the coronagraph. A spectrograph spreads out sunlight into a "spectrum" that displays all the colors of the rainbow. The spectrum of the Sun's surface, for example, is a smooth variation of color, from violet to red, or from short to longer wavelengths. The Sun looks yellow to the eye partly because its yellow

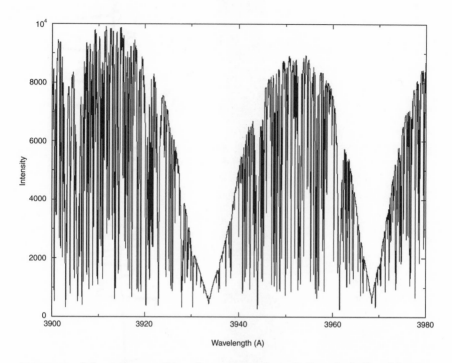

FIG. 1.2   A short segment of the solar spectrum shows the Fraunhofer lines. These are wavelengths where the light is dimmer because of absorption in the solar atmosphere. Wavelength increases toward the right. The two very deep lines are produced by calcium atoms that have lost one electron.

light is strong and partly because the human eye has evolved to be sensitive to this color.

Superposed on the colors are thousands of Fraunhofer lines (fig. 1.2). These lines are wavelengths where the light is somewhat weaker due to the absorption in the Sun's atmosphere by such elements as neon, iron, and calcium.

Every element has its own unique pattern of spectral lines, enabling a physicist to identify it. Atoms of an element absorb and emit light very strongly at the wavelengths of their characteristic lines. In fact, all the terrestrial elements reveal their presence in the Sun by impressing their atomic fingerprints on the smooth, continuous spectrum. By studying the shapes and strengths of these lines, astronomers have been able to determine the chemical composition, temperature, and many other properties of the Sun's atmosphere.

FIG. 1.3 An example of John Evans's "wiggly" lines. The dark ragged image is a photograph of the spectrum, with a dark line center. Distance across the solar disk is displayed vertically. The graph along the spectrum shows the Doppler velocity at each position. The arrows indicate a distance of 10,000 km and 0.45 km/s.

Evans's new spectrograph performed beautifully. On days when the solar image was unusually sharp, he noticed that each spectral line had "wiggles" along the direction of the slit (fig. 1.3). He knew that rising or falling motions of the solar gas would shift the wavelength of a spectral line, because of the Doppler effect (note 1.2). We are all familiar with the rise and fall in the pitch of an ambulance siren as it approaches and passes us. Light behaves in a similar way. An approaching source of light sends us shorter wavelengths (higher "pitch") than it would if it were at rest, and sends longer wavelengths when it is receding.

Evans realized that he could identify each wiggle with a rising or a falling granule that lay along the direction of the spectrograph slit. His new equipment and his excellent site would enable him to study the motions of granules, something that had seldom been done before. He decided to attack the problem.

His method was classical. Early each morning, when the solar image was still sharp, he would expose a series of wiggly-line spectra, made at the center or the edge of the solar disk. Later he would select the exposures that best resolved the granules along the slit. These prizes were then scanned with a machine he had built to determine the precise wavelength shift that the Doppler effect caused in each spectral line within each granule. With a suitable calibration he could convert each shift to an equivalent vertical velocity. It was a slow and tedious procedure, but gradually Evans acquired a set of highly accurate data.

Evans's method had a basic flaw, however. By selecting only the best spectra, regardless of the order in which they were taken, he was discarding (or at least setting aside) some valuable information on the life cycle of granules. Perhaps he planned to investigate this subject later, but his first priority was to compile reliable statistics of granule velocities. In making this decision he was probably influenced by the contemporary idea that convection is a very turbulent process in the Sun. Like the water in a boiling pot, the gas velocities in the Sun vary rapidly from moment to moment and from place to place. The best way to get a handle on granule motions was to measure a great many of them, at all stages of their development, and then to average their speeds appropriately (note 1.3).

Each spectral line in his spectra yielded a different average value of velocity. That was understandable since he knew that the centers of different spectral lines

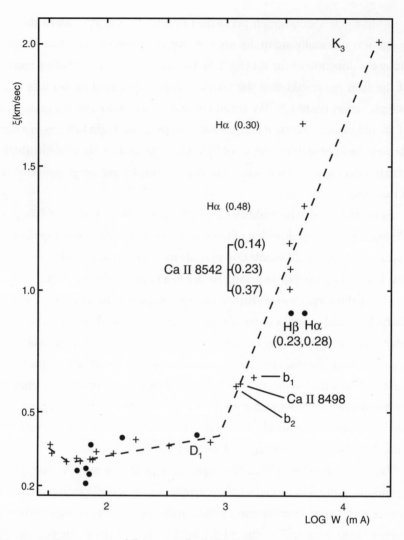

FIG. 1.4 Evans measured the average "turbulent" velocity of the gas at different altitudes in the Sun.

are formed at different heights in the solar atmosphere (note 1.4). For example, a weak line of titanium forms just above the surface, while a very strong line, say of calcium, forms a few hundred kilometers above the photosphere. By observing lines with different strengths, Evans could probe the atmosphere and determine how the vertical gas velocity varies with altitude. Figure 1.4 shows his re-

THE DISCOVERY • 9

sults. On average, the higher one looks, the faster the speed. He took these results to a meeting in Varenna, Italy, in 1960.

LEIGHTON AND HIS METHOD

Meanwhile, at the Mount Wilson Observatory near Pasadena, California, Robert Leighton was also studying motions in the solar atmosphere. Leighton was a professor of physics at the California Institute of Technology, an experimentalist with considerable talent and experience. He was also an excellent teacher and attracted several bright graduate students for doctoral thesis work. With all this he still found the time to write a classic textbook of physics and to edit the famous introductory physics lectures of Nobelist Richard Feynman, a fellow Caltech colleague.

Leighton started out as a cosmic ray physicist, studying, with the help of a large cloud chamber he had built, the decay of mu mesons and the properties of the recently discovered "strange" particles. A fast cosmic ray proton would collide with a nucleus in the chamber to produce a shower of secondary particles. The tracks the particles left in the chamber revealed their masses and electric charges.

In Leighton's day, cosmic rays were the only sources of very energetic protons. Eventually, however, physicists were able to create relativistic particles in the laboratory with powerful accelerators and study their collision products. The importance of cosmic ray research diminished as a result. To gain access to the accelerators a physicist needed to join a large experimental group. Leighton, however, was an individualist who shunned such large operations. So he began to look for greener fields. His cosmic ray observations had been carried out at Mount Wilson Observatory, some 1700 meters above Pasadena, which made him familiar with some of the solar instruments there. He began to think about solar physics.

George Ellery Hale, one of the titans of astronomy, had founded this famous observatory early in the century. In addition to building the largest nighttime telescopes of his day (the 100 inch at Mount Wilson and the 200 inch at Palomar Observatory north of San Diego) Hale made some of the most important discoveries in solar physics. For example, he was the first to detect the strong magnetic fields in sunspots and to chart the systematic changes of their magnetic po-larities through the 11-year activity cycle of the Sun.

Hale built two solar towers at Mount Wilson, one 18 meters tall and a later one 45 meters tall. By raising the solar mirror to such heights, Hale was able to avoid the troubling air turbulence near the ground and to preserve the fine qual-ity of his solar images. To exploit these excellent images, he invented the spec-troheliograph, a compound spectrograph that generates a photograph of the Sun in the light of a single Fraunhofer line (fig. 1.5; note 1.5). Such a photograph shows the small structures in the solar atmosphere at the particular height where the spectral line originates. Fig. 1.6 is an example of a spectroheliogram, made at a specific wavelength in the core of a strong line of hydrogen. It reveals magnetic field lines high in the solar atmosphere. With such photographs, Hale went on to study magnetic fields in so-called active regions, above sunspots.

Leighton conceived a simple modification of the spectroheliograph at the 18 m tower that would record, simultaneously, *two* solar images: one in the light of the blue wing, and the other in the light of the red wing of a single spectral line. Using a clever photographic technique (note 1.6), Leighton was able to subtract the blue-wing image from the red-wing image. The difference in the brightness

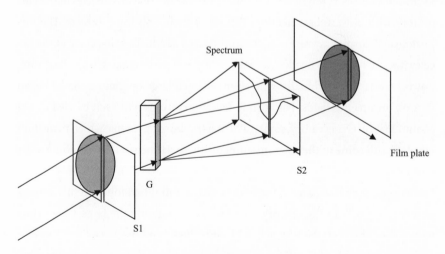

FIG. 1.5   A drawing of a spectroheliograph (see note 1.5).

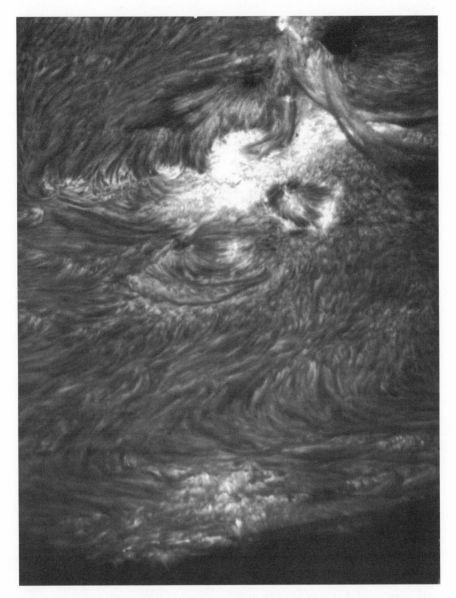

FIG. 1.6 A modern example of a spectroheliogram, an image of the Sun made in the light of a single Fraunhofer line (H alpha) at the red end of the solar spectrum. The image shows the details of the solar chromosphere, which lies above the visible surface. The threadlike forms outline solar magnetic fields.

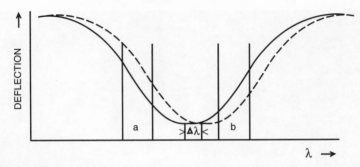

FIG. 1.7   A source at rest emits a spectral line (shown as a solid line), which is centered at a definite, fixed wavelength. If the source recedes from the observer, the whole line shifts to a longer wavelength (to the right) due to the Doppler effect. The shifted line is shown as a dotted line. To determine the shift, one measures the difference of light intensity in two fixed wavelength bands (labeled a and b) that are centered on the undisplaced line. This intensity difference is zero for the solid line (source at rest) but nonzero for the dotted line (source is receding).

of the two wings was a measure of the shift in wavelength, or equivalently, the Doppler velocity at a particular point on the Sun's disk (fig. 1.7; note 1.7). The result was a photograph in which rising and falling blobs were coded as either bright or dark. Such a figure is called a "Dopplergram" for obvious reasons. It has the advantage over a single spectrum of showing velocities over the whole solar disk.

Leighton began to make individual Dopplergrams. From these he was immediately able to make an important discovery: he found that at an altitude of a few hundred kilometers above the visible surface, the Sun was covered in cells in which the gas spreads out horizontally. These cells were much larger than the ordinary granulation (about 30,000 km, or almost three times the Earth's diameter) and lasted for at least several hours. He named these cells "supergranules" and recognized that they must be another aspect of convection in the Sun.

Leighton also saw smaller cells on these single Dopplergrams, some as small as 1700 km, in which the gas was rising or falling vertically at a maximum speed of about 0.4 km/s. Perhaps, he thought, these small ones are associated with granules. If so, he could determine their average lifetime by following the changes in their velocities.

Leighton could have chosen a simple but tedious method. He could have made a long series of Dopplergrams at intervals of, say, a minute, and compared each with the very first one. Each one would have a slightly different pattern of velocity cells. The larger the time difference between Dopplergrams, the larger the difference in the patterns. When he found a Dopplergram that looked sufficiently different from the first one (as measured by a numerical criterion), the time interval between these two would be the desired lifetime. But Leighton chose a more elegant method: he scanned the spectroheliograph slit from one edge of the Sun to the other in five or ten minutes, stopped the machine, changed photographic plates quickly, and then scanned back in the reverse direction. This procedure gave him only two Dopplergrams, but that was enough.

To search for changes in velocity between scans, he subtracted one Dopplergram from the other using his photographic technique and recorded these changes on a new photographic plate. As explained in note 1.8, each point on this plate is associated with a time delay that varies from zero at one edge of the disk to a maximum at the other edge. At the zero delay edge on this plate, he saw no change in the velocities, as expected. Toward the center of the disk, where longer delays occurred between the two original scans, the velocity differences increased, as is expected of a group of granules seen at different times in their lives. Farther toward the opposite edge, at still longer delays, the velocity differences decreased, which was also understandable if the original generation of granules was being replaced by the next generation. The delay time at which this decrease occurred (about 5 minutes) indicated the average lifetime of granules, the very quantity Leighton was looking for. However, still farther across the disk, at even longer delays, the velocity differences *increased* again.

Now, this increase was quite unexpected. Leighton knew that granules emerge randomly over the disk, and indeed the pattern of granules on the disk would change radically as a new generation appeared. Where a granule had been before, none might appear later, after a delay of five or more minutes. He therefore expected that the velocity difference at corresponding points on the two Dopplergrams would remain small for delays longer than the granule lifetime. It did not, however, which raised the suspicion that he was looking at something entirely new.

To check this possibility, Leighton and his student Robert Noyes *added* the two original Dopplergrams to combine the velocity at each point on the disk with the velocity at a later time. In this new plate, the velocity sum vanished after 2.5 minutes. Also, it was easy to see on the Dopplergrams that neither velocity was zero. Since their sum was zero, that must mean that the velocity would be equal and opposite to the velocity at the same point some 2.5 minutes later. The conclusion was clear and startling. *Something on the Sun was oscillating!* In a later observation Leighton recorded three complete cycles of velocity in a spectral line of calcium.

Leighton knew that convection cells don't normally oscillate. His discovery therefore clearly pointed to another kind of phenomenon, something that affects the whole Sun. He told Noyes, "I know what you're going to study for your thesis. The Sun is an oscillator with a period of 300 seconds."

COMPARING RESULTS

In August 1960, Leighton reported his astounding results at the Fourth Conference on Cosmical Gas Dynamics in Varenna, Italy. John Evans was there, gave his own talk, and heard Leighton's. One can only imagine his reaction. On the one hand, he was above all a fair-minded scientist who must have recognized Leighton's outstanding achievement. On the other hand, he must have felt a pang of disappointment in not exploiting his own data fully.

Evans went back to Sacramento Peak and in September 1960 began a new series of wiggly-line observations, this time with a constant time interval between exposures. He invited the aid of Raymond Michard, a skillful French solar astronomer who was spending his sabbatical leave at the observatory. Meanwhile, Leighton enlisted two of his students to extend and refine his preliminary results. Robert Noyes did indeed write a doctoral thesis on the observational properties of solar oscillations, while George Simon wrote one on the properties of supergranulation.

In 1961 Leighton and his students reported their work at a meeting of the International Astronomical Union, a conference of several thousand astronomers held in Berkeley, California. Their talk was a sensation. At the same meeting,

Evans and Michard were able to confirm the Caltech results in complete detail. In addition, almost as a footnote, they revealed a little detail that turned out to be critical.

In contrast to Leighton, Evans could record the oscillations in *several* spectral lines simultaneously. By choosing lines with different strengths, he could track the development of a single oscillation in *altitude* as well as in time. He and Michard discovered that an oscillation often began as a rising motion, with the appearance of a bright granule. Then the "disturbance" would propagate upward like a sound wave for a short time. But within the lifetime of the granule, the entire atmosphere would begin to oscillate like a standing wave. *In short, the disturbance began as a traveling wave and changed within 8.6 seconds into a standing wave* (for a primer on waves, see notes 2.1 and 2.2). This was truly bizarre behavior and nobody knew how to explain it. Evans and Michard proposed a picture in which a granule acts as a piston that sends a wave upward and starts the whole atmosphere bouncing. This picture would mislead theorists for a decade.

WHAT DOES IT ALL MEAN?

In 1962, Leighton and his students published their complete results in a classic paper in the prestigious *Astrophysical Journal*. Later that year the confirmation of Evans and Michard appeared in the same journal. Since all agreed that the piston model could explain the origin of the oscillations, the next most important question was, "What's so special about a five-minute period?"

Leighton suggested several possibilities. The first was that the solar atmosphere is oscillating freely at its resonant period, as a pendulum would. Horace Lamb, the famous English hydrodynamicist, had developed the theory of such oscillations for an isothermal atmosphere as early as 1908. That simple theory predicted 190 seconds, not 300, however. A more complicated model of the Sun might do the trick, but perhaps not.

Second, Leighton considered the opposite alternative that the atmosphere was *not* oscillating freely. Instead, it was filtering out one special sound frequency from the cacophony of noises that the granules produced. But it was difficult to see just how such a thin atmosphere could dissipate all but one frequency.

Among other more speculative schemes, Leighton suggested that the granules might actually oscillate with a period of five minutes. But he was frank to admit that much more work was needed to sort out all the possibilities.

Solar astronomers everywhere were eager to pick up the trail and explore this amazing new phenomenon. In particular, they wanted to determine more accurately the size of the oscillating cells. Leighton had claimed that they have no characteristic size. But other observers did find dominant sizes, some as small as granules (1700 km), some as large as 100,000 km, depending on the precise method they used. Moreover, oscillations with periods as short as four minutes and as long as six minutes were reported. Leighton's clean, simple result was becoming fuzzy. It would take solar scientists another eight years to arrive at a clear picture.

CONFUSION AND CLARIFICATION

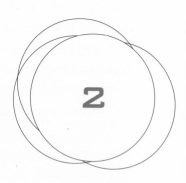

FOLLOWING THE DISCOVERY of the five-minute oscillations, solar astronomers were eager to explore this new phenomenon, but the more they tried, the more confusing the picture became. The problem, as it turned out, was that no one had enough information about the nature of the oscillations to plan for critical observations. As a result, observers would report puzzling and contradictory results.

For example, in 1967 Robert Howard at Mount Wilson Observatory was finding not one dominant oscillation at five minutes but several, a few minutes apart. Worse still, their relative strengths seemed to vary. Furthermore, no two observers could agree on the sizes of the oscillating elements. Some reported 1700 km, others as much as 100,000 km. And the estimated lifetime of an oscillation ranged from 60 to 100 minutes. The oscillations were elusive, a moving target.

Without guidance from the observers, the theorists were also struggling. By 1968, at least three different explanations for the oscillations were competing for attention. The models might be labeled the granule-as-piston, the atmosphere-as-acoustic-filter, and the resonant cavity.

Herman Schmidt and Friedrich Meyer, two scientists at the Max Planck Institute in Munich, Germany, proposed the piston scenario. Schmidt, a slow-spoken, genial man, is the most reasonable and rational person a collaborator might wish for, and Meyer practically quivers with fresh ideas. Together they made a remarkable team.

In 1967 they followed up on John Evans's observation that a new oscillation begins in the photosphere soon after a hot bright granule arrives there from below. They proposed that the granule acts as a piston that slams against the overlying atmosphere and launches a weak shock wave upward. In the wake of this wave, the whole atmosphere begins to bounce up and down (note 2.1 about waves). The two men worked out the consequences of this idea with a numerical model, which was quite successful on the whole. They were able to predict the observed amplitude and period of the oscillations. In addition, they could predict the phase behavior that Evans had observed, a change from a traveling wave to a standing wave (note 2.2). In a paper in the *Zeitschrift für Astrophysik* (1967), they concluded that "individual granules of the sun cause the observed oscillations." Their model had a fatal flaw, however: the predicted lifetime of the oscillation was much too short, even considering its uncertainty. Evans's clue therefore could not be the whole story.

The second type of model (the atmosphere-as-filter) was based on the earlier work of two distinguished British hydrodynamicists, Sir James Lighthill and Sir Horace Lamb. Lamb made important contributions to subjects as different as wave propagation, electrical induction, earthquakes, and the theory of tides. His book *Hydrodynamics* has remained a standard text since 1895. However, the book is highly mathematical. Lighthill once said that one could read all of Lamb's book and never realize that water is wet.

Lamb discovered an essential property of sound waves in a gaseous atmosphere. He proved in 1908 that a gravitating atmosphere, like that of the Earth or Sun, has a critical "cutoff" frequency for sound waves (note 2.3). Only waves with frequencies higher than the cutoff are able to travel freely. At frequencies below the cutoff, the waves are "evanescent"—they fade out a short distance from their source.

Lighthill was also interested in many other subjects, including hurricanes, the flow of water around ships, and the flow of air over aircraft wings. His mind worked at the speed of light. When a speaker at a conference asked, "How does a bacterium swim without legs or a propeller?" Lighthill shot back instantly, "By creating tangential waves on its surface."

When the British civil aircraft industry began to use jets, Lighthill was asked to find a way to reduce the terrific noise the jets made with their exhaust gases. As a result, he developed a theory of how turbulent flow in a gas generates sonic noise. The theory predicted the amount of acoustic power at any frequency.

Derek Moore at the University of Bristol, working with Edward Spiegel at New York University, extended Lamb and Lighthill's results, and in 1964 applied them to the problem of the five-minute oscillations. Pierre Souffrin in Nice, France, had a similar idea a little later. All suggested that the (unknown) turbulent flows in the photosphere would generate sound waves continuously, over a broad band of frequencies. The waves with frequencies above the cutoff would run freely to the top of the atmosphere, while the lower-frequency waves would be reflected. These running low-frequency waves would cause the photosphere to oscillate as a whole, with a period close to the cutoff, or five minutes. In effect, the atmosphere could act as a filter for sound waves, passing the highs and reflecting the lows.

Note, however, that no mention was made of granules. In the Moore-Spiegel scenario, the source of the noise is convective turbulence, not a pistonlike granule, and the source is steady, not impulsive, as in the Meyer-Schmidt model. Nevertheless, in both models the atmosphere reacts similarly. But the Moore-Spiegel model also had some flaws: it did not explain why a single period of five minutes should be favored. Moreover, Evans also found that the oscillations were standing waves, not running waves.

To some other theorists, the strong periodicity near five minutes suggested a *resonance* of some kind. This in turn suggested that a resonating "cavity" exists in the Sun that selects and amplifies a particular frequency just like a flute or an organ pipe. But what and where could it be?

In 1963, Martin Schwarzschild and his colleague John Bahng placed this resonating cavity at the temperature minimum of the solar atmosphere. According to contemporary models of the Sun, the gas temperature falls steadily as one looks higher in the photosphere and then rises steeply in the overlying layer, the chromosphere. Between these two major regions lies a zone of minimum temperature (about 4500 kelvin), bounded on each side by a steep temperature gradient. Sound waves with a period of about five minutes would be trapped indefinitely between these steep gradients, and a standing wave with the same period would be set up, in agreement with Evans's observations of phase.

This model, like the others, had a fatal flaw. The cutoff period at the temperature minimum is actually around *four* minutes, so five-minute waves would be unable to reach that zone from below. At the same time, other theorists, such as Yutaka Uchida in Japan, played with the resonance of buoyancy waves instead of sound waves. But further progress would have to wait for better observations. These were not long in coming.

A CRITICAL CONTRIBUTION

Pierre Mein in France and Edward Frazier in the United States recognized that the best way to sort out the conflicting claims of observers and the competing models of theorists was to obtain data that could be compared to a quantitative theory of waves. This work would require spectroscopic observations over a large area of the solar disk to determine both the wavelength and the frequency of the oscillating elements. Of course Robert Leighton had done something like this already with his spectroheliograph at Mount Wilson, but Frazier and Mein knew they could get more complete data with a time series of spectra.

Frazier made his observations in 1965 at the Kitt Peak National Observatory near Tucson, Arizona. He selected a rectangular area near the center of the disk and stepped its image past the slit of the main spectrograph. A complete scan of the area past the slit took only twenty seconds, and he repeated these scans for fifty-five minutes. At each position of the image on the slit, he photographed a short section of the solar spectrum that contained three suitable spectral lines. One of these (silicon 637.1 nm) forms in the photosphere, another (iron 635.5 nm) forms in the temperature minimum above the photosphere, and the third (iron 636.4 nm) forms between the other two. With this triplet, he could examine the oscillations at three depths in the atmosphere. Observations in white light ("the continuum") sampled the oscillations at the greatest depth.

After the observing session, Frazier measured the Doppler velocity at each

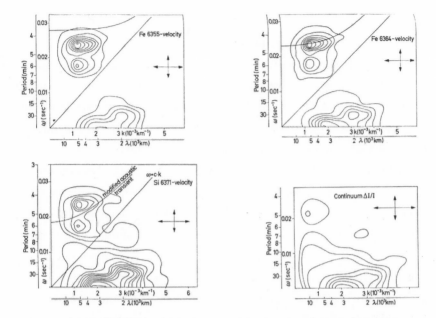

FIG. 2.1   Frazier analyzed his spectroscopic observations and displayed the results in these four diagnostic diagrams. Each diagram refers to a different altitude in the solar atmosphere, with the Fe 635.5 highest and Si 637.1 lowest.

point in his grid, for each moment of time and at three different altitudes. He then carried out a Fourier analysis of the data (note 2.4) to extract the amplitudes, frequencies, and wavelengths of the oscillations. Figure 2.1 shows his results plotted on four "diagnostic" diagrams, one for each spectral line or height in the photosphere and one for the white light. Each diagram shows the "power" or strength of the oscillation at each frequency and horizontal wavelength. The curves on the diagrams provide a way to classify different kinds of waves, hence the term "diagnostic." Such diagrams had been used previously by atmospheric physicists and by oceanographers. To calculate his curves, Frazier assumed the solar atmosphere has a single, uniform temperature.

Hydrodynamic theory predicts that three kinds of waves can coexist in the solar atmosphere (note 2.5). First there are the *propagating sound waves*. They occupy the area above the upper curve in each diagram. These curves define the cutoff frequency for each horizontal wavelength. Only sound waves with frequencies higher than the appropriate cutoff can propagate.

The second kind of wave is the *internal gravity* wave. These kinds of waves are less familiar to most of us, but they do occur in the Earth's atmosphere and in the oceans (note 2.5). They lie below the lower curve in each diagnostic diagram, and this curve is another kind of cutoff. Only gravity waves with frequencies lower than the cutoff can propagate.

Finally, there is the *surface gravity* wave, similar to waves at the ocean surface. It occupies a special diagonal line in the diagnostic diagram. Any wave whose properties place it between the curves in the diagnostic diagram is "evanescent" or nonpropagating. It is a standing wave that fades out a short distance from its source (note 2.2). In Frazier's diagrams, the splotch at the bottom of each diagram lies in the area assigned to gravity waves. Pierre Mein had seen it earlier in his own data, noticed that its periods are longer than ten minutes, and concluded that Leighton's oscillations could not be gravity (or buoyancy) waves, leaving propagating or nonpropagating sound waves as candidates.

Sound waves showed up as double-peaked blobs in the upper left corners of Frazier's diagrams, with periods between 4 and 7 minutes and wavelengths between 3000 and 10,000 km. At all three heights in the photosphere, the stronger peak (corresponding to oscillations with a period of 4.3 minutes) always lies near the acoustic cutoff curve, the upper curve in the diagrams.

Frazier recognized that the cutoff frequency has a special significance, at least in an isothermal atmosphere: it is the *resonant frequency* of the atmosphere. If the atmosphere is excited at that frequency, it will oscillate up and down *as a whole* and build up to a large amplitude. The effect is similar to pushing a child's swing, precisely in phase with its oscillation. He was led to the idea that "the oscillations are standing oscillations excited at the resonant frequency of the photosphere" (*Zeitschrift für Astrophysik*, 1968).

But the photosphere is not isothermal and does not have a unique resonant frequency. Because the temperature decreases at higher altitudes, the cutoff curves Frazier calculated also vary with altitude. As a result, the stronger peak lies *below* the curve at low altitudes and *above* it at high altitudes. This situation conflicts with his idea that resonance occurs exactly at a cutoff frequency. Moreover, he had no explanation for the secondary peak, at around six minutes.

Frazier admitted these were temporary problems, but he insisted his idea was basically correct. As supporting evidence he recalled that he had observed the same oscillatory phase at all altitudes, a unique property of a resonant standing wave. He also argued vigorously against the piston model, pointing out that he never saw the transient waves at low altitudes that the model predicts, and that Evans claimed to see. In his observations, a bright granule would sometimes *disrupt* an oscillation but never *start* one. In fact, many oscillations developed without the aid of a granule.

This development raised the question of where and how the oscillations are excited. Frazier had noticed that not only the vertical velocity but the brightness of the photosphere oscillates. He knew that the light that leaves the photosphere originates at a depth greater than any spectral line he had used, suggesting an origin *below* the photosphere for Leighton's oscillations. Frazier concluded that "the evidence indicates that the oscillations are generated in deeper layers, within the convection zone itself, and by the time that both the oscillations and the convective cells reach the surface, they are relatively uncorrelated with each other" (ibid.). This remark would prove to be vital for the next theoretical advance.

One more piece of the puzzle had to be found before the observational picture was complete: the lifetime of the oscillations. A granule rises, spreads out horizontally, and then sinks, all in about ten minutes. The oscillation Meyer and Schmidt predicted would last at most a few minutes longer. Therefore, when Georges Gonzi and François Roddier found that a typical oscillation could last for an hour or more, the final nail was driven into the coffin of the piston model. Such a long lifetime argues for a resonance effect.

A THEORY AT LAST

The stage was now set for a critical interpretation of the observations. Not surprisingly, the same interpretation came from two independent groups, at nearly

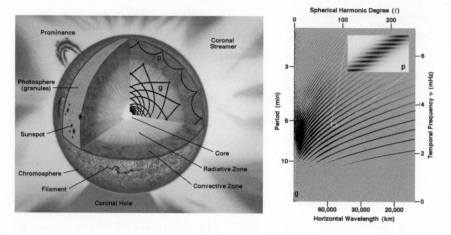

FIG. 2.2   A cutaway view of the Sun, showing its core, radiative zone, convective zone, and photosphere.

the same time—something that happens frequently in science when the fog of confusion finally begins to lift.

In 1970, Roger Ulrich, an assistant professor of astronomy at the University of California at Los Angeles, published a short paper entitled "The Five Minute Oscillations at the Solar Surface." (In a sense, the title misses the main point of the article, which is that the observed oscillations originate *below* the surface.) Ulrich was well equipped to grapple with the problem. As an undergraduate he studied chemistry at Berkeley, but for graduate work he switched to astronomy. Louis Henyey, an inspiring professor, piqued his interest in the internal structure and evolution of stars. For his doctoral thesis Ulrich developed an improved theory of convection in cool stars like the Sun.

In such stars, hydrogen is converted to helium in a central core by a chain of thermonuclear reactions. The energy released by the conversion flows outward as radiation. At some point along the way, the star can no longer transport the energy by radiation alone, so its gas begins to churn as it *convects* heat to the surface. Hot blobs of gas are buoyant, so they rise to the surface, where they cool by radiating to space, and then sink. In the Sun, these motions occur in a *convection zone*, estimated to be about a quarter of a radius thick, or about 175,000 km (fig. 2.2).

When Frazier's paper appeared, with its bold suggestion that the oscillations originated below the photosphere in the convection zone, Ulrich was immediately interested, because the problem would then fall into his area of expertise. Just two years after receiving his doctoral degree, he published a theory of oscillations that explained many of their confusing aspects. Most important, his theory gave specific predictions that could be tested with observations.

Moore and Spiegel had already shown how turbulent motions in the convection zone would generate sound waves with a broad range of frequencies that could propagate in all directions. Ulrich set out to learn how and where some of these waves might be trapped. For convenience he used an existing theory of wave propagation, an extension of Lamb's theory published earlier by William Whitaker.

Ulrich discovered that sound waves with certain characteristics could be trapped between two horizontal reflecting boundaries. The upper boundary for these waves lies just below the top of the convection zone. Above this boundary, the gas temperature drops off rapidly and therefore the cutoff frequency rises. Waves with frequencies below the cutoff frequency are partially reflected, back into the convection zone.

The lower boundary lies at a depth (a few tens of thousands of kilometers) that depends on the frequency and the horizontal wavelength of an incident wave. At this boundary the increasing temperature and speed of sound act to bend a low frequency wave back toward the direction from which it came (see note 2.6 on refraction). The two boundaries form a resonance cavity in the convection zone that traps a low frequency wave in a standing wave pattern. (If you blow across the top of a beer bottle, you can get a similar effect. The bottle is the resonant cavity, and the noise you hear is the sound wave that your turbulent breath excites in the bottle.) In the Sun, the resonant cavities are not perfect. Some wave energy leaks out of the top and into the overlying photosphere. Here the waves are evanescent, or nonpropagating. They can force the photosphere to oscillate at the cutoff frequency at the upper boundary.

All these conclusions followed directly from Whitaker's theory. To make concrete predictions, however, Ulrich needed a definite temperature model of the solar convection zone. This was no problem for him since he had already developed the ideal tool for the purpose, his doctoral thesis. Once he had a model, he was

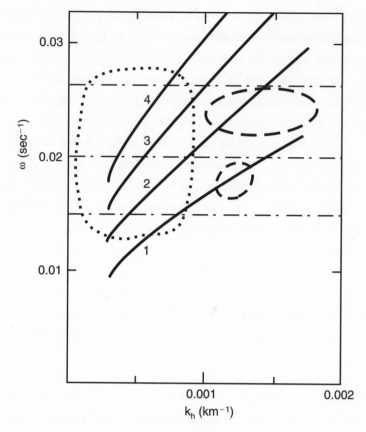

FIG. 2.3 Ulrich predicted that waves trapped in the solar convection zone could have only particular discrete combinations of frequency and horizontal wavelength. These combinations lie on separated curves in this diagnostic diagram. The integers on the curves indicate the number of nodes (plus one) of the standing wave between the reflecting layers.

able to calculate the actual wavelengths and frequencies of the standing and evanescent waves.

Figure 2.3 shows his results, plotted on a diagnostic diagram. Only those waves with particular combinations of frequency and horizontal wavelength can be trapped, and these combinations fall on *discrete curves* in the diagram, like pearls on a string. More precisely, half the vertical wavelength of a wave must fit between the reflecting boundaries an integral number of times before it can form a standing wave. In Ulrich's diagram, the different curves are distinguished by precisely this integer. We can also see on the diagram the outlines of the observations of several observers, including Frazier's double peak.

Why hadn't any observer seen this pattern of lines in an empirical diagnostic diagram? Ulrich explained that nobody had yet observed a sufficiently large area of the Sun's disk for a sufficiently long time. As a result, the observations lacked the resolution in space and in time to see the curves (or "ridges") in the diagram. Ulrich specified the minimum observing requirements: an area 60,000 km in diameter and a time of at least one hour.

Now the observers could pick up the trail once again. They could either prove or disprove Ulrich's predictions.

PARALLEL LINES MEET . . .

While Ulrich was working, he was unaware that two competitors were snapping at his heels. John Leibacher was just finishing up his doctoral thesis at Harvard. Robert Stein, another Harvard graduate, was now an instructor at Brandeis University. Both had worked around the edges of the oscillation problem: Stein had studied the generation of sound waves by convection by extending Lighthill's theory to the conditions in the sun, concluding that most of the acoustic noise had periods between thirty and sixty seconds, far too short to explain five-minute oscillations; Leibacher had investigated wave propagation in solar-type atmospheres for his doctoral dissertation.

These young theorists stumbled on the idea of trapped waves while studying the behavior of a single acoustic pulse, a solar clap of thunder that contained a broad range of frequencies. Later, they went through the same chain of reasoning that Ulrich had followed. They recognized that turbulence in the convection zone would generate acoustic noise continuously and that some of this sound would be reflected below the photosphere by Lamb's cutoff phenomenon. Unlike Ulrich, they were uncertain about the nature of the lower reflecting boundary, but their numerical simulations showed that many types of boundaries could work. With an upper and a lower boundary they had a resonant cavity, and the rest was easy. Their numerical simulations showed how the photosphere was driven up and down as a whole by the evanescent waves that leaked through the upper boundary.

Their short note, "A New Description of the Solar Five-Minute Oscillation," was accepted for publication in October 1970 but was not actually published until after Ulrich's paper appeared in December of that year. Nevertheless all three scientists are credited with cracking the most intriguing problem in solar physics of recent times.

In 1972, Charles Wolff added some finishing touches to this model of trapped sound waves. He was struck by the fact that a single oscillating cell could be as large as a tenth of a solar radius, or 70,000 km. Ulrich's theory, he noted, applied strictly to a *slablike* atmosphere in which the temperature, for example, is constant on infinite horizontal planes. A model that ignores the curvature of the Sun would be limited to oscillation wavelengths much shorter than a solar radius. So Wolff proceeded to work out the properties of standing waves in a realistic convection zone. In this model, the zone is a spherical shell and the solar radius acts as a fundamental scale length. Therefore, the vertical wavelength of an oscillation is constrained to be a simple fraction (such as 1/3, 1/23, or 1/40) of the solar radius.

In 1975, Hiroyasu Ando and Yoji Osaki, a team working at Tokyo University, extended this basic three-dimensional model. They recognized that because the upper reflecting boundary lies close to the Sun's surface, a sound wave could lose energy by radiating to empty space, introducing a phase shift between changes of temperature and velocity in a wave. They included this effect in more accurate predictions of the observable frequencies and wavelengths.

CONFIRMATION

We now had an attractive explanation for Leighton's five-minute oscillations. As the saying goes, however, the test of the pudding is in the eating. The history of science is littered with beautiful theories that don't quite match reality. Unless someone could actually detect Ulrich's ridges, the trapped wave model was just a pretty toy.

Ulrich gave the task of testing his theory to his graduate student, Edward

Rhodes. Today, Rhodes is a professor at the University of Southern California and is recognized as an expert in the field of helioseismology. But in 1974 he was a lean and hungry doctoral candidate, eager to complete a thesis.

Rhodes traveled to Sacramento Peak Observatory to do his work. At the time, this observatory had some of the most advanced instrumentation in the world. An evacuated solar tower, the first ever built, produced the sharpest solar images ever seen, and a huge spectrograph displayed Fraunhofer lines in exquisite detail. But the jewel of the observatory was a linear array of 128 photoelectric diode detectors, capable of recording a spectrum digitally. At this time, before the invention of two-dimensional electronic detectors (charged couple detectors, or CCDs), the observatory's array was unique. All of this equipment had been designed by a young genius, Richard Dunn, who developed into the most innovative instrumentalist of his time. His inventions enabled astronomers like Edward Rhodes to make outstanding discoveries.

George Simon, a staff scientist at the observatory, assisted Rhodes in using all this unfamiliar equipment. Rhodes scanned a square region at the center of the Sun with the diode array, recording the Doppler shift at each point. For several days he repeated his scans at a steady cadence for over four hours. By January 1975 he had all the data he needed for his thesis and began the arduous task of analysis, laboring all through the summer of 1975.

But then, in November 1975, Rhodes got a rude shock: he had been scooped. Franz-Ludwig Deubner, a staff scientist at the Fraunhofer Institute in Freiburg, Germany, had been working independently of Rhodes and had published his results first.

Deubner, now a professor at the University of Würzburg, is a big, cheerful man with a huge appetite for work. After Ulrich published his recipes for critical observations in 1970, Deubner began to try them out at the Fraunhofer Institute's telescope on the idyllic isle of Anacapri. This strange-looking refracting telescope had a radical design: it lacked a protective dome. The idea was to avoid the turbulence that builds up as sunlight heats a dome during the day, thereby preserving the good quality of the solar images. Evidently, the scheme worked.

Figure 2.4 shows Deubner's diagnostic diagram, compared with theoretical curves computed by Ando and Osaki. The comparison speaks for itself: Deubner had proved that Ulrich, Leibacher, and Stein had hit on the correct explanation for Leighton's five-minute oscillations.

Rhodes completed his thesis, then published a preliminary report in 1976 and a complete paper (with Ulrich and Simon) in 1977. He was also able to confirm Ulrich's theory, and made two important discoveries in addition. First, a slight mismatch between the predicted and observed ridges (fig. 2.5) showed that the thickness of the convection zone that was commonly assumed in solar models (about a quarter of a solar radius) was too small; in fact, the zone might be as thick as 0.38 of a radius. (The latest measurements indicate a fractional radius of

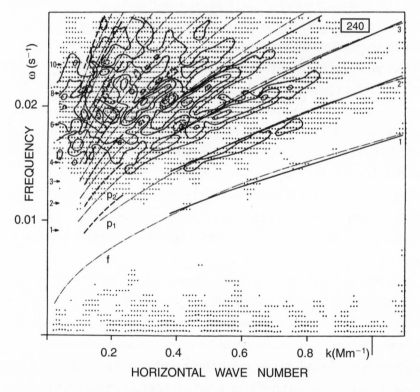

FIG. 2.4 In 1975, Franz-Ludwig Deubner finally succeeded in detecting the curves (or "ridges") in the diagnostic diagram that Ulrich had predicted. His observations (the contours) were slightly displaced from the curves, which suggested that the solar model needed improvements.

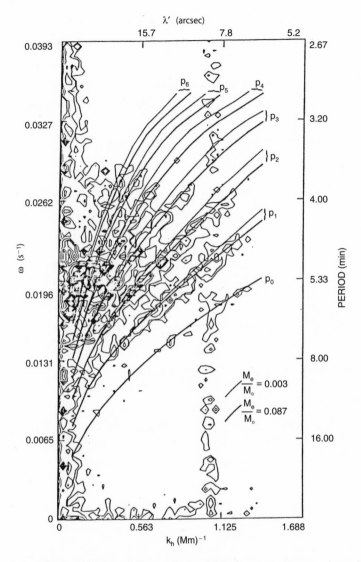

FIG. 2.5   Rhodes, Ulrich, and Simon obtained this diagnostic diagram about the same time as Deubner. It also confirms Ulrich's theory of the solar oscillations.

0.291.) And second, Rhodes realized that the oscillations might be used as probes of the internal rotation of the Sun. We are sure to hear much more about both of these discoveries.

With the observational proof of the global oscillation theory, a new discipline in solar physics was born: helioseismology. From here on, progress would be rapid.

A CLOSER LOOK AT
SOLAR OSCILLATIONS

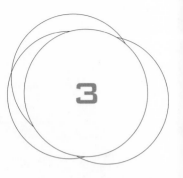

3

MOST OF US are familiar with musical instruments such as guitars, flutes, and drums. Each instrument produces sound by vibrating in its own way. The standing waves (see note 2.2) on a guitar string and in the air column inside a flute are one-dimensional, while the waves on a drumhead are two-dimensional; but those in the Sun (and Earth) are three-dimensional. Each added dimension provides many new ways a vibrating object can oscillate, and the Sun is no exception. Solar oscillations are complicated. What do they actually look like? Let's take some time to examine them.

Figure 3.1 shows a recent snapshot of the five-minute oscillations at the Sun's surface. These are the kind of Doppler velocity maps that astronomers record every minute or so, for days or months, in order to tease out the characteristics of the internal oscillations of the Sun and, from these, to derive important physical properties. But as you can see, the map is practically featureless, a monotonous sea of tiny bright and dark patches. Each patch oscillates up and down with its own period for a few cycles and then fades away, to be replaced nearby with another patch. Without a theory of oscillations to guide them, astronomers could make no sense of such maps.

Fortunately a realistic theory exists due to the efforts of many astronomers during the past century to understand the variability of stars. By studying the

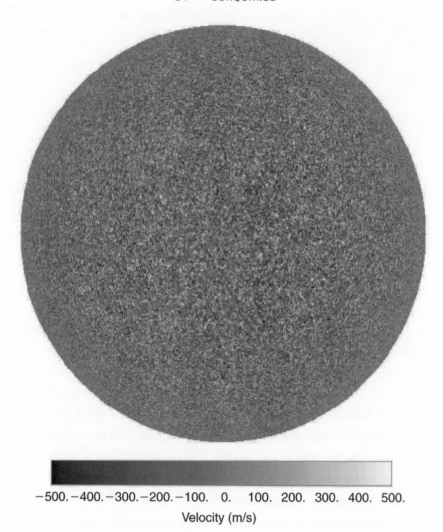

−500. −400. −300. −200. −100. 0. 100. 200. 300. 400. 500.
Velocity (m/s)

FIG. 3.1   A snapshot of the Doppler velocity of the gas at the surface of the Sun. Dark points are sinking, bright points are rising. The average speed in either direction is about 100 m/s.

pulsations of certain kinds of stars, they have been able to probe their interiors, at least to some extent. In a sense, the Sun is only the latest example of this ongoing work.

As early as 1879, German physicist August Ritter suggested that stars vary in brightness because they pulsate periodically. Later on, Sir Arthur Eddington, the eminent British astrophysicist, constructed the physical theory of stellar pulsation. In 1917 he applied his theory to the famous Cepheid variables, hot stars that pulsate in brightness with periods between one and fifty days. He explained that these stars pulsate in purely *radial* expansions and contractions, like a balloon, and release stored heat during their cycle. His theory predicted a relation between a Cepheid's period and its intrinsic brightness, a connection that allowed Harlow Shapley and Edwin Hubble to use Cepheids to measure extragalactic distances.

Thomas G. Cowling, another brilliant Briton, was one of the first to study *non-radial* pulsations of a model star. In such stars, the surface is wrinkled into complex periodic spatial patterns as sound waves bounce around in the stellar interior. As we shall see, this is precisely the situation in the Sun. In 1949, Cowling and his colleague R. A. Newing added *rotation* to their pulsation models, an even closer approach to the real Sun. Paul Ledoux, the great Belgian theorist, developed nonradial pulsation theory even further and in 1951 applied it to the star Beta Canis Majoris, which lies close to Sirius, the brightest star in the northern sky.

These pioneers were followed by many others, including John Cox and Arthur Cox in the United States; Douglas Gough in Cambridge, England; Hiroyasu Ando and Yoji Osaki in Japan; and, more recently, Jørgen Christensen-Dalsgaard in Denmark. At the present time the basic theory for the Sun is highly developed but slowly evolving as new helioseismic observations are analyzed.

Before looking at the theory, let's explore some far simpler systems.

MUSIC TO MY EARS: NODES AND MODES

Pluck a guitar string and listen for a moment. The string sings in a dominant note, with some "overtones" that give the guitar its individual character. These notes correspond to standing waves on the string, with different wavelengths, as

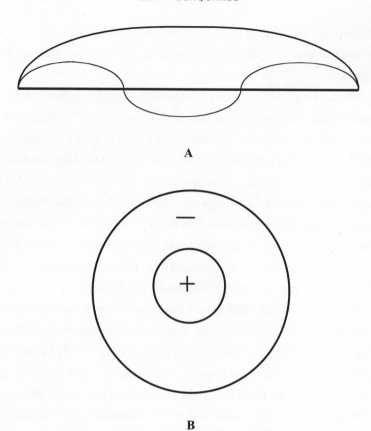

A

B

FIG. 3.2 Standing waves on a string (A) and a drumhead (B). On the string, half-wavelengths must fit precisely between the nodes. On the drumhead, one simple overtone pattern is shown. Here the nodes lie on circles.

figure 3.2A shows. The dominant or fundamental note corresponds to the longest wavelength, half of which extends between the two fixed end points or nodes. The overtones have wavelengths that are fractions ($\frac{1}{2}$, $\frac{1}{3}$, $\frac{2}{3}$, etc.) of the fundamental's and have stationary nodes along the length of the string. Each of these patterns is called a vibration "mode."

We could distinguish among these different modes by labeling them with the number (say, N) of nodes they have along the length of the string. The fundamental would have N = 0, the first harmonic (with half the fundamental's wavelength) N = 1, and so on.

Next, imagine that we hit a drumhead exactly at its center. The blow could excite a variety of modes. In the simplest one, the center of the drumhead vibrates up and down while the circular edges are fixed. We see no nodes along a radius of the drum; so we could label this mode as N = 0. However, if we strike the drum harder, a circular pattern as in figure 3.2B could appear. We would hear a definite change in the sound as higher overtones are excited. In fact, one interesting question in physics and mathematics is, "Can you hear the shape of a drum?"

A drumhead's vibration varies in two dimensions. Nodal *lines* separate the parts of the drumhead that move in opposite directions. To describe such an oscillation, we'd need two numbers, one (say, N) to count the number of nodal lines along a radius and a second number (say, M) to count the number of nodal lines around a circle. In figure 3.2B, N = 1 and M = 0. Obviously, many more complicated modes are possible, as both N and M are varied in different combinations.

Finally, imagine that you are sitting in a rectangular concert hall and the orchestra is tuning up by playing a middle A note, which has a frequency of 440 cycles and (at room temperature) a wavelength of 0.8 meters. If this hall had a length of 80 meters, a width of 40 meters, and a height of 24 meters, it would be poorly designed. It would form a perfect *resonance cavity* for the A note, and the echo would be deafening. Because its dimensions are simple multiples of the note's wavelength, a standing wave would fill all three dimensions of the hall as the orchestra tunes. There would be two hundred nodal *planes* in midair, ranging along the length of the hall, one hundred across its width and sixty from floor to ceiling. To describe this standing wave, we would need three numbers, let's say N = 200, L = 100, and M = 60.

SOLAR MODES

Because the Sun is also three-dimensional, it also requires three numbers to label one of its modes of vibration. Astronomers have agreed on their names. The number of nodes along a radius, N, is called the *radial order;* the number of nodes around the equator at the surface, M, is called the *azimuthal order;* and the number from pole to pole, L, is called the *angular degree.*

In figure 3.3 we see how the Sun's modes change shape at the surface as L and

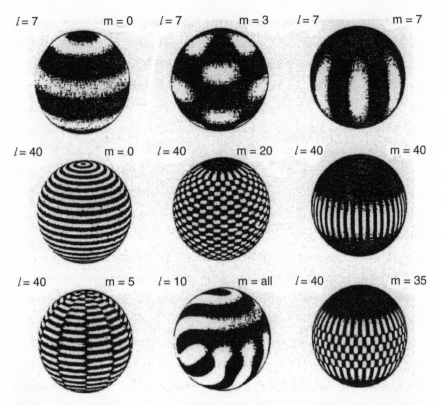

FIG. 3.3 Oscillating modes on the Sun's surface are distinguished by two numbers, L and M. The degree L specifies the number of nodes from pole to pole, the azimuthal order M specifies the number of nodes around the equator.

M vary. White regions are rising in this snapshot, and dark regions are falling. All the regions within a mode oscillate together (or "in phase") at the same frequency. If M = 0 (see the two images on the upper left of the diagram), the nodal lines that separate rising and falling areas follow the parallels of latitude, and the number of parallel lines is fixed by the degree L. If L is fixed and M is increased (as in the top row), the shape of the mode changes from latitudinal bands to meridional bands. And when both L and M are large but different, as in the lower-right image, the mode has a complicated checkerboard shape.

Figure 3.4 is a cutaway view, showing the oscillation nodes inside the Sun for a mode with $N = 14$, $L = 20$, and $M = 16$. This mode, like all others, has a unique frequency, which in this case is 2935.88 microhertz (a period of 340.61 seconds).

The real Sun (fig. 3.1) looks significantly different from these patterns. The reason is that many modes are present at the same time and overlap at the surface. Each mode vibrates up and down at only a few centimeters per second, but when a large number of them piles up at any one point, their sum can reach a few hundred meters per second, which is easily detectable.

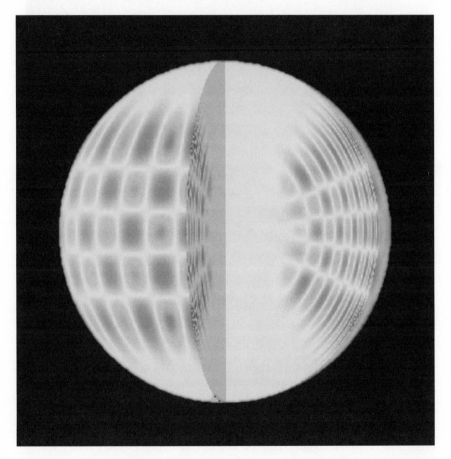

FIG. 3.4 If we could remove a slice of the Sun we could see the nodes inside it, spaced along a solar radius. In reality each interior node is a spherical surface.

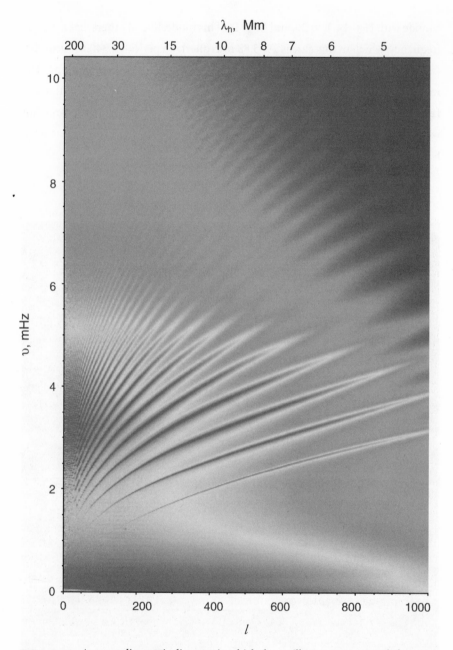

FIG. 3.5 A recent diagnostic diagram, in which the oscillatory power at each frequency and degree (L number) is displayed.

We can estimate the total number of modes present at any time in the following way. In the latest diagnostic diagrams (for example, fig. 3.5), at least twenty ridges have been observed. They correspond to the radial orders, $N = 1$, 2, 3, etc. For each value of N, values of L as high as 500 have been observed, and for each L there are twice as many M's. So the number of modes present in the Sun at any one time can be as large as $20 \times 500 \times 1000$, or about *ten million*. They all overlap, in time and space. As a result, a snapshot of the Sun's disk (fig. 3.1) looks featureless, like coarse sandpaper.

With so many modes present at one time, how can a bewildered observer unscramble them? How does she hear the shape of the solar drum? In other words, how does she progress from snapshots like figure 3.1 to diagnostic diagrams like figure 3.5? For a quick answer, see note 3.1.

A TOY SUN: NODES AND MODES

The oscillations we see at the surface of the Sun are reflections (literally) of the standing sound waves that fill the interior. Each standing wave (or N, L, M mode) is trapped between the surface and some critical depth, just as the wave on a guitar string is trapped between its ends. That critical depth (or "turning radius") will depend on how the velocity of sound varies with depth into the Sun. So, in order to make sense of the observations and use them to probe conditions inside the Sun, we first need to know how the sound speed varies. How do we learn that? One way is to calculate a numerical model of the Sun. (Later on we will discuss other methods.)

A model consists of a set of tables that list the temperature, density, pressure, composition, and other properties of the solar gas at every depth or radius. Building such a model is no easy task. To be satisfactory—that is, realistic—the model must reproduce the observed luminosity (total energy output) and radius of the Sun at its present age, requiring a deep understanding of nuclear physics, energy transport by convection and radiation, and hydrodynamics. With all that in hand, the model is completely determined by only four governing equations and three critical parameters from observations: the solar mass, initial chemical composition, and age. Note 3.2 sketches the way such a model is constructed.

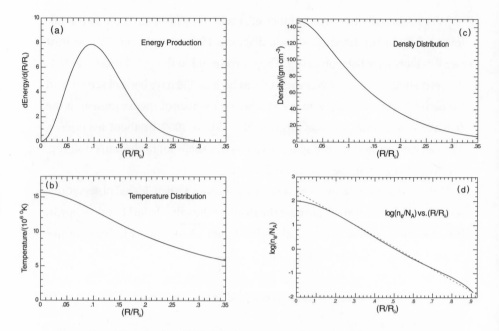

FIG. 3.6   A "standard model" of the Sun, computed by John Bahcall and Roger Ulrich. The panels show the radial distributions of (a) energy production, (b) temperature, (c) gas density.

Through many trials, and much computation, astrophysicists have arrived at a set of working models that differ only slightly in details. No one model has been adopted as "The Standard," but all of them are based on the simplest possible physical assumptions and the most accurate physical data. (John Bahcall, an astrophysicist at the Institute for Advanced Study in Princeton and an expert on solar neutrinos, has generated dozens of *nonstandard* models to try to account for the puzzling solar neutrino observations.) Each model must pass the minimum observational test of matching the Sun's radius and luminosity at its present age of 4.6 billion years. Figure 3.6 illustrates one such model.

The next step is, in effect, to jiggle this toy Sun. We would introduce a small "perturbation" in all the physical quantities in the governing equations and ask what conditions must be satisfied among the perturbations for stable oscillatory motions to exist. We would find three kinds of waves that turn out to be relevant to the observations so far. (Other types of waves are possible and may be relevant in the future.) These are *sound waves, internal gravity waves,* and *surface gravity waves,* in which pressure, buoyancy, and gravity, respectively, supply the necessary restoring forces.

We saw in chapter 2 that traveling sound waves can be trapped inside the Sun because they are reflected at certain depths. This is also true of internal gravity waves. The theory tells us that three critical frequencies (which vary along a radius) determine where in the Sun a sound wave or gravity wave of a given frequency will be reflected (fig. 3.7). These are the *buoyancy frequency,* the *Lamb fre-*

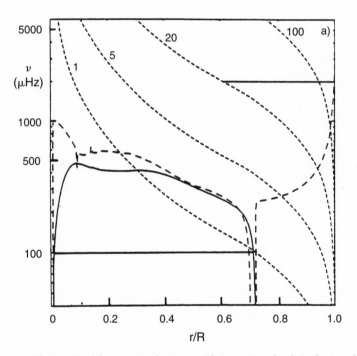

FIG. 3.7   Three critical frequencies that vary with increasing depth in the Sun determine the regions where sound waves and gravity waves are trapped: buoyancy (solid line), Lamb (small dashes with L values), and cutoff (large dashes). The heavy straight line anchored on 100 microhertz indicates a gravity wave trapped between the center and a fractional radius of about 0.7. The heavy straight line on the top right indicates a sound wave with frequency of 2000 microhertz and degree L = 20.

quency, and *Lamb's cutoff frequency* (f_{co}), which we've met in note 2.3. (See note 3.3 for more details.)

A trapping region has the shape of a spherical shell that acts as a resonant cavity for sound waves with a unique frequency and direction, like the concert hall we mentioned earlier. Only the traveling waves whose wavelength fits neatly into the region, in all three dimensions, survive to form a standing wave pattern or mode. There are p-modes for sound waves, g-modes for internal gravity waves, and f-modes for surface gravity waves. As noted earlier, each mode can be labeled by the number of nodes (N, L, M) it has in each of the three dimensions within the Sun.

TURN AND TURN AGAIN

We have been rather vague about the way in which sound waves travel inside the Sun. Figure 3.8 may help to clarify the situation. Here we see two waves looping inside the Sun. Each wave travels down into the Sun at an oblique angle after bouncing at the surface. Because the speed of sound increases inward, each wave front is refracted and loops back toward the surface.

In general, the depth at which a wave turns back depends on its frequency and degree or L number. Only waves with the longest horizontal wavelength (that is, with L = 0, 1, 2, 3) can reach the deep core of the Sun. As we shall see, this result forces astronomers to design special equipment to probe the core.

WHO IS KNOCKING AT THE DOOR?

So far we've avoided the question of how the oscillations are excited—for a very good reason: nobody knows for certain. The most likely answer is that turbulent motions of the gas in the convection zone generate traveling sound waves of all frequencies (recall James Lighthill's work). Only waves that fit neatly into the Sun's natural resonant cavities survive as the standing waves that we detect at the surface. That still leaves a huge range of discrete frequencies from which to create the surface patterns.

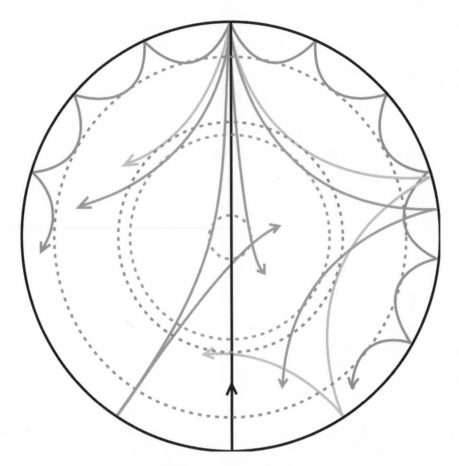

FIG. 3.8 Sound waves of different frequencies and "degrees" (L) take different paths through the Sun. Waves with low L (say 1, 2, 3, etc.) travel large distances horizontally in broad arcs and penetrate deeply into the Sun. Waves with large L (say 100 or more) follow sharply curved ray paths and return to the surface without probing deeply into the Sun.

In this respect, the Sun is both similar and dissimilar to the Earth. Everyone knows that a major earthquake sets off traveling waves in the Earth. We tend to forget that many tiny earthquakes occur every day, keeping the Earth trembling constantly at a low level. The same things happen on the Sun: in July 1996, for example, the SOHO satellite detected a "sunquake" following a major explosion on the Sun. But the gentler turbulent motions in the convection zone keep the Sun ringing like a bell continuously. At least, that is our understanding today.

· · · · ·

As you read the following chapters you may wish to refer back to this one to recall the unfamiliar notation used to distinguish oscillation modes. With that suggestion in mind, let's move on.

THE SCRAMBLE FOR
OBSERVATIONS,
1975 – 1985

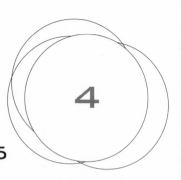

THE NEWS spread like wildfire, first through the community of professional astronomers and then through the newspapers: Franz-Ludwig Deubner had peered into the heart of the Sun and confirmed Roger Ulrich's prediction that the Sun was ringing like a bell. The implications for astronomy in general and solar physics in particular were enormous. Now astronomers could apply the same techniques for probing the Sun that seismologists used for probing the Earth. They would soon know whether their carefully constructed picture of the solar interior was correct. And even in these early days, the prospect of probing other stars beckoned.

Nothing succeeds like success, they say. Solar physicists around the world decided to jump into this new field of helioseismology. Observers tried out several different technologies, each with its own advantages and all contributing to the rapidly developing picture. Theorists were proposing new models to fit the data.

Indeed, from the very first this new field was characterized by close collaboration between observers and theorists. The incredible precision of the new observations drove the theorists to build ever more realistic models. At the same time, the theorists took a skeptical look at some of the claims made by observers. A fast-paced dialogue developed in the pages of *Nature,* a science journal that publishes new results within a couple of weeks.

As new data accumulated, three large questions were debated. First, how does the temperature vary inside the Sun? Unless we could determine the central temperature more precisely, we couldn't account for the deficit in the number of solar neutrinos (see note 4.1). Second, how does the interior of the Sun rotate? In particular how does the core rotate? Third, what exactly is the present composition of the Sun? We knew that hydrogen is most abundant, but the proportions of helium and heavy metals were still uncertain.

Looking back, we can see that most of the new helioseismologists chose one of two basic approaches to observing the oscillations. Either they scanned an image of the Sun, acquiring data at each point across the disk, or they recorded the oscillations in "integrated" sunlight, treating the Sun like a pointlike star.

THE FIRST PEEK INSIDE

After Edward Rhodes and Franz-Ludwig Deubner published their independent confirmations of Ulrich's theory, they joined forces to exploit the new phenomenon of solar oscillations. In 1977, they returned to Sacramento Peak and obtained new data with more care and better observing conditions and equipment. Now they were able to resolve as many as eight ridges in the diagnostic diagram with much higher precision than ever before. From their data they were able to derive for the first time the variation of rotation speeds *below the visible surface* of the Sun.

Their result was based on a simple but clever idea of Rhodes. Sound waves that travel in a flowing gas are speeded up or slowed down, depending on which direction they take. The situation is similar to a motor boat on a river. The boat's progress, relative to the shore, is faster when it moves with the flow than when it moves against it.

Each standing wave in the convection zone can be resolved into two waves traveling in opposite directions (see note 2.2). One wave travels in the direction of the Sun's rotation (east to west) and is speeded up, relative to the observer on Earth. The other wave is slowed down correspondingly. Therefore, the observer sees two wave frequencies that differ by twice the angular speed of rotation of the gas in which the waves are traveling. From the difference in frequency one can determine the speed of rotation. The next step is to estimate, from a solar model,

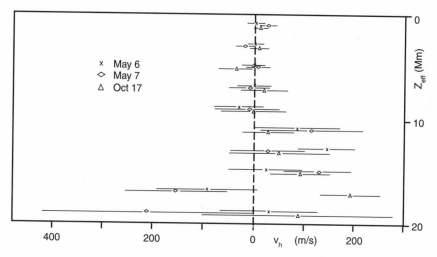

FIG. 4.1   Deubner, Rhodes, and Ulrich were the first to measure the speed of rotation under the Sun's surface.

the depth at which this standing wave is trapped. That will depend on the frequency and horizontal wavelength (or degree L) of the wave mode.

Following such a procedure, Deubner and his colleagues determined the variation of rotation speed down to a depth of about 15,000 km below the photosphere. Figure 4.1 shows their result. At a depth of about 8000 km, the speed is larger than the surface speed (2 km/s) by about 100 m/s. That was an interesting result, because they knew that sunspots rotate about 100 m/s faster than the surrounding surface. Perhaps the magnetic fields of sunspots are anchored in this layer, 8000 km deep, and are dragged through the overlying gas.

Note how the error bars grow larger with increasing depth. Below 15,000 km the three independent sets of data are uncertain and discordant. Nevertheless, this first estimate of the rotation of the interior was a triumph for this new field of research.

LOOKING AT THE WHOLE SUN AT ONCE

Solar astronomers are always complaining that the Sun is too dim. That may seem surprising because to the eye, at least, the Sun is the brightest object in the sky. But to study velocity oscillations in the classic way I have described, observers are forced to spread out the light into its spectrum, thereby reducing its intensity. Then, in order to determine the horizontal wavelengths of the waves, they need to resolve patches no larger than a few thousand kilometers. Because each patch emits only a tiny fraction of the Sun's total output, the amount received decreases even further. And because the oscillation periods are limited to a few minutes, an observer cannot afford to collect light indefinitely. So in the end, one may literally run out of light.

The Sun is doing its level best, however, so the observer must compromise some requirements in order to gain in other respects. This is the strategy that led several groups to develop a technology that would yield more precise velocity measurements by sacrificing all spatial information. Instead of forming a solar image and recording the light from each small area, they chose to pass sunlight from the whole disk directly into their instruments, thereby gaining enormously in light.

The instruments these groups developed are based on the resonant scattering of light in a vapor of such metals as sodium or potassium (note 4.2). A metallic vapor scatters light only in very narrow bands around discrete spectral lines. These bands can be shifted slightly in wavelength by embedding the vapor in a magnetic field. By suitable means, a detector can switch between bands in each wing of a spectral line. In this way, the Doppler shift of the line in full-disk sunlight can be determined.

When recorded in this way, the short-wavelength oscillations, which have many peaks and valleys over the solar disk, tend to cancel one another; only oscillations that cover a whole hemisphere or quadrant survive. These simple modes of low degree ($L = 0, 1, 2$) are especially interesting because they penetrate deep into the sun, but they are difficult to detect because their amplitudes are only a few meters per second. (Recall that the many overlapping five-minute modes combine to produce an amplitude of a few *hundred* m/s.) Nevertheless, these resonance scattering cells are so stable and so sensitive that they can easily measure such small speeds.

Two groups were especially interested in such an instrument. George Isaak at the University of Birmingham in England had the idea as early as 1961. In 1973 he and his colleagues built cells with either sodium or potassium as the vapor and began to observe the Sun soon afterward. At the University of Nice in France, François Roddier and his student Eric Fossat had a working model by 1971. These efforts gained momentum just as helioseismology was taking flight.

Eric Fossat and his colleague Gerard Ricort were the first to report scientific results. They had observed at several locations, including Nice itself, over a period of three years and found oscillation periods as long as ten minutes.

The real surprises came, however, from the group at Birmingham. Because Birmingham's climate is unsuitable for solar astronomy, James Brookes, George Isaak, and Henry van der Raay set up their instrument in September 1974 at the Pic-du-Midi Observatory in the French Pyrenees. This famous solar observatory is located on a narrow ridge at 2877 meters, with a spectacular view into Spain. Visitors approach the fortresslike buildings in a swaying cable car, a really chastening experience. The ridge has an awesome serpentlike formation of light rock embedded in its front face, which contributes to one's impression of approaching a witch's castle.

The weather was foul. Only one day out of twelve yielded usable observations. Nevertheless, Brookes and company found four strong oscillations in that data; surprisingly, a five-minute period was not among them. Instead, they found periods of 29, 40, 58, and 160 minutes, with amplitudes as large as 3 m/s. They identified the 58-minute period with the fundamental radial pulsation mode of the sun and the 29-minute period as the first harmonic. So these were apparently long-period p-modes, similar to the usual five-minute oscillations but offering the advantage of sampling the deepest parts of the Sun.

No contemporary model of the Sun could account for the 160-minute period, however. The Birmingham group recognized that only a homogeneous sun, without a hot core, could pulsate at 160 minutes. That idea was absolute nonsense, of course. Such a model would not even predict the correct luminosity for the Sun, the most basic requirement of a model.

Perhaps the 160-minute period was not a sound wave after all, but an internal gravity wave, an entirely different species, in which buoyancy rather than pressure is the relevant force. The problem with that explanation was that gravity waves were thought to be trapped deep within the Sun and unable to

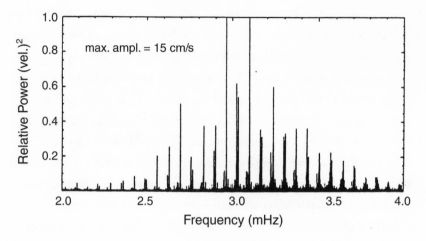

FIG. 4.2 The spectrum of low-degree solar oscillations, a group of discrete frequencies spaced around 3 millihertz or a period of five minutes, with nearly uniform spacing.

reach the surface with any detectable strength (see fig. 3.7). So no reasonable explanation seemed possible. We will return to this curious puzzle later in this chapter.

The Birmingham group continued to push ahead with its potassium cell. They changed their technique in order to sample the classical five-minute oscillations, and in 1979 Antoine Claverie and friends reported observations made over three years at Izana, Tenerife. They had also observed simultaneously from two stations, Izana and the Pic-du-Midi, for a year. Altogether, they had accumulated over five hundred hours of data, some with continuous stretches of fifteen hours.

Why was so much data necessary? The answer is simple: the longer you count the swings of a pendulum, the smaller the error you make in estimating its period. The same holds true for the Sun: longer continuous observations yield more precise oscillation frequencies.

Their strenuous efforts paid off. Their oscillation spectrum (shown in fig. 4.2) consists of a series of sharp peaks, centered at a frequency of 3 millihertz (a period of five minutes) and separated by precisely 67.8 microhertz. The tallest peaks correspond to an amplitude of only 15 cm/s.

The group identified these peaks by using three slightly different solar models. These had been computed by Jørgen Christensen-Dalsgaard and Douglas Gough, the two brilliant theorists at Cambridge University whom we will meet over and over again. The peaks turned out to be p-modes corresponding to L = 3, 4, and 5 and N = 16 to 28. The best fit with the 67.8 microhertz spacing was obtained with a model Sun that began life with less helium and heavy metal than a "standard" model (19% by mass instead of 25% for helium, 0.4% instead of 2% for the metals).

George Isaak got very excited about these results. In a paper in *Nature*, he suggested that similar observations could be made of other stars, thereby opening a whole new field of research. (As we shall see in chapter 11, he was absolutely correct!) He also pointed out that a lower fraction of heavy metals could account for the observed deficit of solar neutrinos (note 4.1). That was the good news. But he also worried that the low proportion of helium (relative to hydrogen) detected in the Sun conflicted with the established theory of the creation of all cosmic helium in the Big Bang (note 4.3).

Christensen-Dalsgaard and Gough bounced right back with a rebuttal. Although they couldn't be absolutely sure, the Sun's initial helium content could have been the standard value, 25% by mass. In constructing the models used by the Birmingham group, they had neglected the influence of the solar atmosphere. For sound waves with frequencies well below the Lamb cutoff (note 3.3), that is a valid assumption. But waves with frequencies close to the cutoff (as in the Birmingham observations) are reflected so close to the Sun's visible surface that the inertia of the atmosphere introduces a change in frequency. When these changes were taken into account in a model with a standard proportion of helium, the predicted frequencies matched the Birmingham observations nicely.

Bear in mind that this whole discussion revolved about uncertainties of frequency of only 1.5 microhertz or one part in two thousand!

NEWS FROM ARIZONA

In 1975, the same year in which Deubner confirmed Ulrich's explanation of the five-minute oscillations, a professor of physics from the University of Arizona stunned the audience at a conference in Paris. Henry Hill announced the discovery of at least a dozen oscillation periods, ranging from seven to fifty-two minutes, accurate to about 5%.

Christensen-Dalsgaard and Gough immediately offered a menu of possible oscillation modes that might correspond to these new periods. The most likely candidates were standing sound waves of low degree (L = 0, 2, or 4). Some predicted p-modes were missing in the data but the agreement with the rest was impressive. This was another example of the predictive power of existing solar models.

These two theorists also proposed a new method of probing the Sun. The usual "forward" method was to calculate a grid of models with slightly different compositions and to predict the frequencies that should be observed. The best match with observations would then select the best model. In their alternative method, they used the fact that each mode has maximum strength only over a limited range of depths, which one could estimate with a relatively crude model. By combining observations of several modes that overlap at a chosen depth, they could determine a correction to the assumed sound speed at that depth. In this way, it might be possible to "invert" the data, to improve solar models, just as seismologists had been doing routinely. This was a bold proposal, which the observers grasped immediately.

THE SHAPE OF THE SUN

The Arizona group hadn't used a resonance scattering cell, a spectrograph, or any other conventional device to detect these oscillations. In fact, Hill and his students, Timothy Brown and Robin Stebbins, were not primarily interested in solar oscillations, but in testing Albert Einstein's theory of general relativity.

As one test of his theory, Einstein attacked one of the outstanding problems in celestial mechanics. The orbit of Mercury is an ellipse that pivots about the Sun by 5600 arc-seconds per century. Newtonian theory had failed to predict the exact rate of precession, missing the mark by 43 arc-seconds per century, even after including the gravitational effects of all the other planets. Einstein's theory hit the mark exactly.

Einstein offered a second test of his theory. He predicted that light from a distant star would be deflected from a straight line to Earth by the intense gravity of the sun. If the light just grazed the edge of the Sun on its way to Earth, it should be deflected by exactly 1.76 arc-seconds. In 1919, two British expeditions observed a similar effect during a total eclipse of the Sun, but poor observing conditions yielded a deflection uncertain by perhaps 20%. In the following fifty years, many attempts were made to improve the accuracy of this eclipse experiment, but without achieving perfect agreement with Einstein's prediction (note 4.4). Nevertheless, it was clear to all that Einstein's predictions were essentially confirmed, and so the general theory of relativity was established as one of the great landmarks of the twentieth century.

Robert Dicke, an associate professor at Princeton University at the time, was not convinced, however. Dicke was a superb experimental physicist who had very high standards of proof. He wrote that an uncertainty in a measurement of, say, 10%, was not good enough to exclude alternative explanations of the same event.

Suppose the Sun were not the perfect sphere Einstein had assumed. Suppose instead that it was slightly flattened (or "oblate"), perhaps because of its rotation. After all, a spinning ball of gas should develop a bulge around its middle because of centrifugal forces. It would only take a flattening of a few parts in a hundred thousand to account for the rate of precession of Mercury's orbit. Such a small oblateness was nearly impossible to measure.

Nevertheless, Dicke and his students did measure it. The problem was trying to measure the Sun's polar and equatorial diameters accurately enough. Turbulence in the Earth's atmosphere commonly distorts the edge of the Sun's image by a few arc-seconds, while the effect they sought was no larger than a few *thousandths* of an arc-second. They built a special telescope and scanned the outer edge of the solar image with a rotating mask. Over several years they gradually eliminated dozens of instrumental and atmospheric sources of error. After analyzing many hundreds of hours of data, they concluded that the Sun was flattened more than enough to dispute Einstein's prediction. Their announcement provoked an uproar in scientific circles that lasted for over a decade.

But let's get back to Henry Hill.

GRAVITY WAVES?

Hill had been a close colleague of Dicke's in the development of the Princeton telescope. For various reasons he decided to move off to Arizona, build an improved version of the oblateness telescope, and repeat Dicke's experiment.

To determine the Sun's oblateness, Hill and his students measured the Sun's diameter at its equator and poles, and compared the two diameters. To overcome the effect of atmospheric turbulence, their equipment isolated a small segment of the extreme edge of the disk and scanned back and forth across the edge. They observed for many hours, over several years, gradually improving their technique and their equipment. In 1974 they announced that they found that the Sun was no more flattened than its surface rotation would require. This result flatly contradicted Dicke's.

Dicke defended his own careful work with the suggestion that the core of the Sun was spinning much faster than its surface. After much more debate, the issue was settled in Hill's (and Einstein's) favor. However, Dicke's rebuttal had opened up the whole question of the rotation of the solar interior.

As a by-product of their oblateness analysis, Hill and his students discovered that the diameter of the Sun oscillates with periods as short as 6.5 minutes and as long as 66 minutes. As we mentioned earlier, Christensen-Dalsgaard and Gough were able to identify most of these periods with standing sound wave patterns of low degree. Any period longer than about 63 minutes could conceivably belong to an internal gravity wave, however.

Hill was intrigued with the possibility that his technique could detect the long-sought nonradial gravity waves, the Holy Grail of helioseismology. These waves extend into the solar core and could, in principle, yield unique information about the core's temperature, rotation, and chemical composition. So he and his student Thomas Caudell reanalyzed some of their data and found two candidate periods, 45 and 66 minutes. In 1979 they presented arguments that these were indeed gravity waves with L = 20 to 40.

Then, in 1983 Hill and his student Randall Bos announced a set of periods ranging up to two hours, some of which persisted without a break for forty-one days. They suggested that these could also be gravity waves. Hill's claims were criticized on many counts, principally that his data were too noisy to make reliable identifications, but he vigorously defended his methods, his analysis, and his conclusions.

Hill was not the first, nor the last to claim the prize for gravity waves. Phillip Scherrer, for example, at the Stanford Solar Observatory had also claimed to detect them. Both ran into a storm of criticism. Observers asked why some gravity waves were present in the data while other waves, equally likely to be present, were not. Theorists harped on the basic question of how a gravity wave, trapped deep inside the Sun, could ever influence the surface unless the convection zone was unbelievably thin. These long-period oscillations remained controversial for two decades.

THE ROTATION OF THE CORE

Meanwhile, the Birmingham group in England was making news again with their resonance cells. They obtained twenty-eight days of data in 1980 at Teide Peak on Tenerife, one of the Canary Islands, and found thirty-three discrete peaks in their oscillation spectrum. They now had enough continuous data to resolve the *splitting* of oscillation frequencies.

As figure 4.3 shows, the frequency of an oscillation mode splits into several closely spaced components because of the nonuniform rotation of the interior of the Sun. The number of components depends upon its degree, L. For example, the frequency of the L = 0 mode does not split at all, the L = 1 frequency splits into a triplet, the L = 2 frequency into a quintet, and so on. The frequency separation of the components is just the rotation frequency of the shell in which the mode is trapped, typically about one microhertz. This is just what Antoine Claverie and friends found in their data. (I will defer a physical explanation for the splitting to chapter 5.)

From the separation of the frequency components, Claverie and colleagues could derive an estimate of the rotation period of the core. A core half the size of the Sun would be rotating at three times the equatorial rate at the surface. If the core were as small as a seventh the size of the Sun, it could be spinning *nine times as fast* as the surface. These were startling results and they sup-

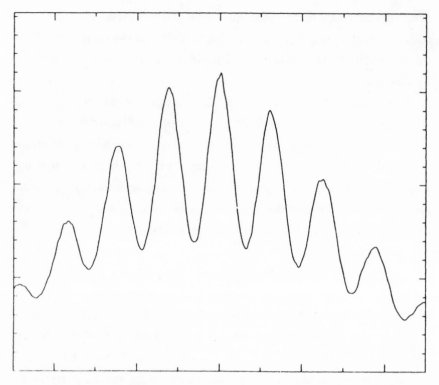

FIG. 4.3   An example of the splitting of an oscillation frequency into components, because of the internal rotation of the sun. This pattern corresponds to L = 3, and reading
from left to right, M = −3, −2, −1, 0, 1, 2, 3. The separation of the peaks corresponds to
the Sun's average rotation frequency.

ported Dicke's conjecture. However, no combination of core size and speed
could account for Dicke's large oblateness. Who was right? What was the core
doing?

A year later, the Birmingham group revised their estimates. They had observed
from Hawaii and Izana simultaneously for more than seventy days. From the
splitting of the low L peaks they discovered a rotation period of 13.1 days, or
about half the surface equatorial period. With this improved data they could as-
sert that the core of the Sun rotates at least twice as fast as the surface. Two
decades would pass, however, before astronomers would agree on the rate of rotation of the core.

NEWS FROM THE SOUTH POLE

Longer is better. Everyone knew that to obtain the most precise oscillation one had to observe as many cycles as possible without any interruptions. Both the Birmingham and Nice groups had lengthened their observing days by using two stations separated in longitude. But the stations were not far enough apart to see the Sun continuously, so that unavoidable gaps still remained in the data. Such gaps introduce spurious peaks in the oscillation spectrum that are difficult to remove after the fact.

Gerard Grec, Eric Fossat, and Martin Pomerantz found a better solution. They decided to observe from the Amundsen-Scott Station at the south geographic pole in Antarctica. In the austral summer, the Sun remains above the horizon for six months. With luck and good weather, it might be possible to accumulate several days of uninterrupted observations. Historical records suggested that clear, windless weather might hold for as long as five days, a tempting target for any red-blooded solar astronomer. So in the austral summer of 1979–1980, these three intrepid pioneers set up their equipment at the South Pole to observe the five-minute oscillations with a sodium resonance cell.

At the time, Pomerantz was the director of the Bartol Research Foundation, a private institution situated at the University of Delaware. A cosmic ray physicist, he had launched hundreds of high-altitude balloons so his instruments could escape the blanketing atmosphere. Eventually he tried conducting his experiments at the South Pole. Because of the pole's altitude (about 3000 meters) and frigid temperatures (dipping below minus 20° F in the summer), ground level there resembles the stratosphere elsewhere, making it ideal for cosmic ray research. Pomerantz became a strong advocate of the pole for all kinds of research, including stellar, infrared astronomy, and, eventually, helioseismology.

By 1979, Grec and Fossat were already seasoned helioseismologists. They had developed their sodium vapor cell at the University of Nice and had observed with limited success at several locations several years earlier. For the polar expedition they modified the cell to cope with the cold, while Pomerantz built a simple vertical telescope at Bartol.

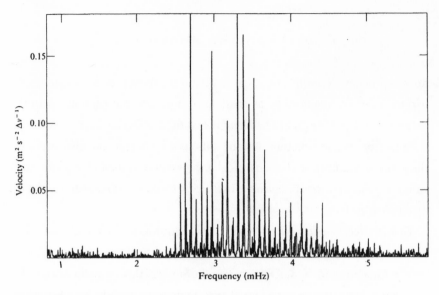

FIG. 4.4 An oscillation spectrum similar to that of figure 4.2, but recorded at the South Pole with less background noise.

To avoid the plume of heat rising from the station, the three moved their base some 8 km upwind. A well-insulated laboratory (a box two meters by three) was installed in a deep trench, covered with a plywood roof, and buried in deep snow. The telescope and cell were located thirty meters downwind to avoid the heat and vibration of their diesel generator.

They worked hard and they were lucky. They gathered five full days of uninterrupted data, then extended the time to seven days with a few hours of cloud. Figure 4.4 shows their oscillation spectrum, in which the noise level is lower by a factor of ten than any attained previously. They could detect oscillations with amplitudes as small as *four centimeters per second.*

Like the Birmingham group, they saw a row of peaks separated by precisely 68 microhertz, which confirmed that they had detected p-modes with L = 0 to 3. Their data were so good that they could resolve the *width* of the main peaks, which yielded a minimum lifetime (about two days) for these oscillations. These beautiful results would give the model builders more meat to chew on.

NEWS FROM SPACE

In 1983, Martin Woodard and Hugh Hudson, two solar physicists at the University of California at San Diego, announced in the pages of *Nature* that they had discovered five-minute oscillations in the amount of sunlight that reaches the top of the Earth's atmosphere, the "irradiance." This was a new way to observe the oscillations, and the result surprised most astronomers. We used to think of the Sun's output as the "solar constant." If the Sun's energy output varies at all, we thought, it should vary slowly, perhaps in step with the solar cycle (eleven years), perhaps over centuries (note 4.5).

Whether the irradiance varies and whether such a variation could affect the climate on Earth were important questions that had arisen in the first decade of the twentieth century. Samuel Langley, the director of the Smithsonian Astrophysical Observatory, suspected that the answers to both questions were "yes." He died before he could reach any definite conclusions, but his successor, Charles Greeley Abbott, spent the following thirty years trying to determine whether the Sun's irradiance varies and by how much.

Abbott made extensive measurements, first from sea level, later from high mountains, and eventually from balloons. By 1920, he thought he had determined real variations of a few percent, which correlated with changes in weather patterns. However, the Earth's atmosphere absorbs sunlight by different amounts, depending upon the wavelength and the presence of dust and moisture. Abbott's critics charged that his corrections for atmospheric absorption were not accurate enough. He continued his efforts through the 1930s, adding more observing stations and improving his equipment, but was never able to satisfy his critics.

The only way to eliminate the large and variable corrections for atmospheric absorption was to carry a sensitive instrument into space. During the 1960s and 1970s, several geophysics satellites were equipped with sensitive instruments (pyroheliometers) to monitor the irradiance, but these early instruments were not sufficiently stable to detect changes smaller than a percent over a year or more. This situation changed with the development of the Active Cavity Radiometer Irradiance Monitor, or ACRIM, by Richard Willson at the Jet Propulsion Labo-

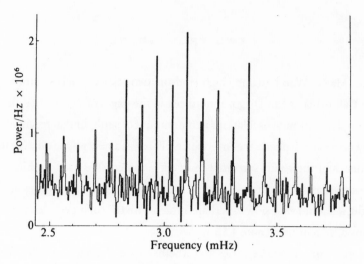

FIG. 4.5   The five-minute oscillations as recorded in full sunlight by the ACRIM instrument aboard the Solar Maximum Mission.

ratory in Pasadena, California. The ACRIM could measure the solar constant with an absolute accuracy of 0.1%, and measure variations over short periods with even higher accuracy. A copy of the ACRIM was carried aboard a solar satellite (the Solar Maximum Mission) in 1980 and delivered ten months of superb data before the satellite lost control of its pointing.

Woodard and Hudson determined the frequencies with which the irradiance varies (fig. 4.5). A familiar family of spikes appears at discrete frequencies centered on 3.1 millihertz (or a period of 5.4 minutes). Note that the strongest peaks correspond to amplitudes of only *two to four parts per million* in the irradiance, a practically infinitesimal signal but absolutely real.

From the spacing of the peaks, Woodard and Hudson determined that the oscillations corresponded to the L = 0, 1, and 2 modes, with radial orders N from 19 to 24. But, although the individual peaks were broadened, there was no sign of the splitting that the Birmingham group had seen. This was a discrepancy they had to explain. Perhaps their data were too noisy or some slight instrumental effects had crept into them.

In any case, the width of the L = 0 peak, which rotation doesn't split, yielded a lifetime of the oscillation of at least two days and possibly six. Imagine a bell that would continue ringing for two days! The Sun is evidently quite free of any kind of friction that might damp out the oscillations.

A TALE OF TWO CITIES

We now return to the story of the infamous 160-minute period. It all began in the old Soviet Union.

Most parts of the USSR were too cloudy to be able to observe the stars very often. The only exception was the Crimea, in the southern Ukraine. There the Soviets built a major facility, the Crimean Astrophysical Observatory. During the 1950s they equipped it with everything needed to explore the universe, and provided it with a large staff. The lord of this huge complex was Andrei Severny, a genial slow-spoken astronomer who had the political skills to prosper in the bureaucratic wars of the Soviet Union.

Severny was best known for his work on the magnetic fields of the Sun. He and his engineers developed an instrument to measure the weaker fields (a magnetograph), much like the one at Mount Wilson Observatory. With this device he was able to investigate the large-scale fields of the Sun as well as the stronger fields in sunspots. When he read about Robert Leighton's discovery of the five-minute oscillations, he was intrigued and decided that his observatory should enter this field of research. Valeri Kotov, one of his colleagues, came up with a clever scheme for measuring oscillations (note 4.6). They built the device and began to observe around 1974.

By 1976, Severny, Kotov, and T. T. Tsap had enough data to arrive at a surprising conclusion: the Sun oscillates with a period of 160 minutes. They pointed out that if their result were confirmed, it would require a major overhaul of conventional solar models. All these models contained a hot core where thermonuclear reactions generated the Sun's energy. In such a model no standing sound wave should have a period longer than about an hour. But a model without a hot core, a very young Sun, would oscillate at 160 minutes.

Naturally, the immediate reaction of solar physicists around the world was skepticism. In their view, a single observation was inadequate to discard contemporary models of the Sun, which had been painstakingly refined and which accounted for so much, so well. There had to be another explanation for the Crimean result.

The easiest explanation was that the period was an artifact of Severny's data analysis, not a real solar oscillation. Observers familiar with such problems pointed out that 160 minutes is exactly one-ninth of a day. Perhaps this odd interval had crept unnoticed into Severny's analysis. Another explanation was that Severny's observations were corrupted by oscillations of the Earth's atmosphere. Perhaps as he looked through the atmosphere toward the Sun, the intervening air imposed a spurious oscillation of 160 minutes.

Severny's result would soon have been forgotten if, in the same year, 1976, the Birmingham group had not confirmed it. They, of course, used an entirely different instrument at a different time and location, and analyzed their data with particular care. Could the long period be a real solar signal despite the reservations of the theorists?

Severny and his colleagues were encouraged to continue to observe. By 1978 they had four years of data in hand. They then claimed that the 160-minute oscillation had continued smoothly, without a single break in all that time. The Sun was apparently like a reliable clock that ticked steadily for years. This behavior was totally unlike the five-minute modes, which change phase abruptly after a few days. The problem with Severny's claim was that, like all solar observatories, Crimea had to shut down at night. With a daily break in their observations, Severny and friends couldn't be absolutely sure that the 160-minute oscillation continued without a break.

The plot thickened after the Stanford Solar Observatory entered the picture. John Wilcox, the founder of the observatory, was a physicist who had discovered in 1962 that the extension of the Sun's magnetic field into interplanetary space has a "sector" structure, in which the polarity of the field alternates with solar longitude. Because the interplanetary field connects to the Earth's field and guides disturbances from Sun to Earth, Wilcox wanted to be able to predict the field. For this purpose, daily measurements of the large-scale solar magnetic field were needed and the Stanford Observatory was built for that purpose in mind.

Around 1978, Phillip Scherrer, a student of Wilcox, had the bright idea of collaborating with the Crimean Observatory. Each observatory could observe when the other was in darkness, and together they could acquire an unbroken string of data, subject only to the vagaries of weather. By 1979, they were able to state in the pages of *Nature* that the Sun had ticked with a 160-minute period without a break throughout their joint study. No oscillation of the Earth's atmosphere was likely to behave in this way, so the signal was almost definitely solar.

Grec and Fossat also saw the 160-minute period in five days of their South Pole data, but they were more cautious. It might still be noise, they said tentatively.

There remained the nagging suspicion that somehow, despite the best efforts of Crimea and Stanford, a spurious one-ninth of a day had remained buried in the data. The two groups plodded on for another year. By that time they had enough data to declare that the period was not 160 minutes, but actually 160.01 minutes! That, they thought, should answer their critics. The signal *had* to be solar.

Scientists are open-minded but intellectually conservative. As Thomas Kuhn wrote in his landmark book, *The Structure of Scientific Revolutions* (1970), only overwhelming evidence can overthrow an established paradigm. To most solar physicists, Severny's claims were not persuasive because they were incompatible with established models of the sun. If the 160-minute oscillation was a gravity wave, for example, why had no one seen other gravity waves?

Kotov and his colleagues at the Crimean Observatory continued to pursue the 160-minute oscillation all through the 1980s but never managed to convince the world that it is a real solar signal, much less a gravity wave. Kotov eventually claimed to find the 160.01 period in special galaxies and even a quasar. The issue gradually faded from public attention, however.

• • • • •

In summary, the first decade of oscillation research had revealed a rapidly rotating core, a normal abundance of helium, and a deeply satisfying agreement between the predictions of standard solar models and increasingly precise observations. Much remained to be done, however. The rotation of the interior was still very uncertain, for example, and the nagging question remained: "Where have all the neutrinos gone?" Some changes in a solar model could conceivably account for the neutrino deficit, but would these changes square with the oscillation frequencies? Only time would tell.

WHEELS WITHIN WHEELS

5

THE SUN'S
INTERNAL ROTATION

FIVE BILLION YEARS AGO, a slowly rotating interstellar cloud of molecular hydrogen collapsed to form a proto-Sun. As this blob continued to contract under its own gravity, it spun faster and faster, just as a skater spins faster when she raises her arms above her head. In some simulations of this collapse, the dense core of the Sun retains most of the spin of the original cloud, while the outer shells rotate more slowly. The Sun is, after all, just a big ball of gas. Gravity holds it together, but its different layers are free to slide over one another because the friction ("viscosity") of the gas is rather small.

Over a period of five billion years, though, even a small amount of friction might be enough to smooth out differences in rotation speed. Perhaps the interior is actually rotating as a solid body would. As recently as 1980, we couldn't be sure. We could see the Sun's surface, of course, and here the rotation speeds were obvious. They had been measured repeatedly, two different ways, by tracking such long-lived tracers as sunspots, or by measuring Doppler velocities across the disk. Figure 5.1 shows the results from the two methods, which agree quite well. The surface rotates, relative to the Earth, with a period of twenty-seven days at the equator and more like thirty-five days at the poles. Astronomers call this behavior "differential rotation in latitude."

So, the Sun doesn't rotate like a cannonball, at least near the surface, but how

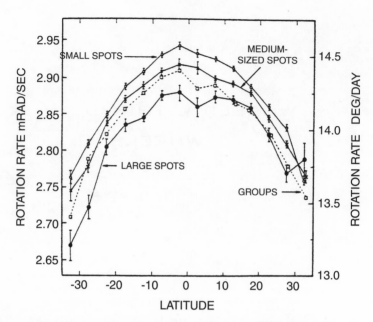

FIG. 5.1 Sunspots can be used as tracers of the rotation of the Sun. The results show that the surface of the Sun rotates faster at the equator than at the poles.

does the Sun's deep interior spin? Some general rules in hydrodynamic theory suggested that the Sun's angular speed is constant on cylindrical shells (see fig. 5.2) and decreases inward. The most sophisticated computer simulations, incorporating the best theory of convection, confirmed that concept beautifully and were able to reproduce the differential rotation at the surface. In figure 5.3 we see the output of one of these simulations. Gary Glatzmaier, one of the foremost experts on solar convection and rotation, used the state-of-the-art computers at the National Center for Atmospheric Research to produce this satisfying result in 1987.

Helioseismologists were planning to test this prediction: throughout the 1980s, at least six experimental groups were competing to map the Sun's internal rotation, once and for all time. They used different types of equipment and different strategies, but all of them were trying to measure the *splitting* of oscillation frequencies.

Sound waves, as we learned in chapter 4, are carried along in a moving medium. Like a jogger on a carousel, a wave traveling in the same direction as the Sun's rotation sweeps by an observer on Earth faster than one that travels in the opposite direction. As a result, the observer sees the frequency of a wave shift up or down, depending on the direction of the wave. For the rotating Sun this means that waves moving eastward inside the Sun have slightly lower frequencies than waves running westward. Since each mode (standing sound wave) is composed of pairs of waves running in opposite directions, rotation "splits" the mode frequency into two or more parts. From the amount of frequency splitting, an astronomer can calculate the rotation speed. (To learn more about frequency splitting, read note 5.1.)

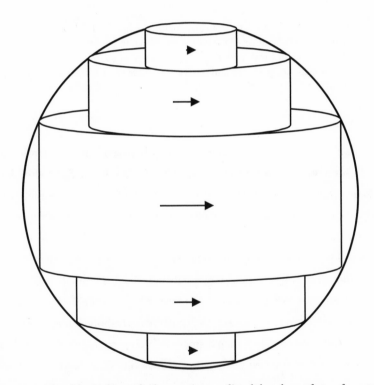

FIG. 5.2   The old paradigm of solar rotation predicted that the surfaces of constant angular speed were cylinders, centered on the Sun's axis. The outer cylinders were supposed to rotate faster than the inner ones.

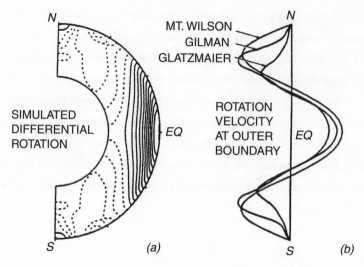

N

MT. WILSON
GILMAN
GLATZMAIER

SIMULATED
DIFFERENTIAL
ROTATION

EQ

N

ROTATION
VELOCITY
AT OUTER
BOUNDARY

EQ

S   *(a)*

S   *(b)*

FIG. 5.3   A numerical simulation of solar rotation, which included the most complete physics available in 1987, confirmed the paradigm shown in figure 5.2. The contours in (a) are the predicted cylindrical surfaces of constant angular speed in the interior. The curves in (b) compare the predicted and observed latitude variations of speed at the surface of the Sun.

There's more. Different modes are trapped between different depths in the Sun. Modes with a low degree (L) penetrate deeply into the Sun, while high-L modes are trapped close to the surface. Thus, by combining the modes judiciously, one can assign a depth to each rotation speed, or, in other words, one can build up a map of internal rotation (note 5.2 describes this process of "inverting" oscillation frequencies to map the interior).

The amount of splitting caused by rotation is devilishly small. A five-minute period corresponds to a frequency of 3333 microhertz, while the rotation period (say twenty-seven days) that causes the splitting corresponds to only 0.4 microhertz. The only way to detect such a small difference in frequency is to observe continuously for a sufficiently long time. How long? Take the reciprocal of 0.4 microhertz and you get thirty days. During that time your equipment must not wander in frequency by more than the splitting you are trying to measure. This is the complicated problem these investigators faced.

Actually, I have overstated the problem somewhat. It turns out that the width of the frequency splitting pattern increases as the degree (L-value) of the mode increases. So, for example, a mode with L = 30 will have a splitting pattern thirty times wider than that of a mode with L = 1 and would require only one day to be detected. But there's a catch. A mode with L = 30 won't touch the core. Only the low L-modes do, and to measure their splitting accurately requires many days.

In fact, the structure and rotation of the core have only recently been tied down satisfactorily, so we will postpone a look at the core until chapter 9. In this chapter we will follow each of four competing groups to see how a new picture of solar rotation outside the core emerged during the 1980s. We begin with two clever chaps at the Kitt Peak National Observatory in Tucson, Arizona.

PROBING THE SUN'S WAISTLINE

The McMath-Pierce Solar Telescope is the largest in the world, with a mirror 1.5 meters in diameter. The telescope stands atop Kitt Peak in the Tohono Indian reservation, 90 km southwest of Tucson. As one approaches, one can easily pick out the distinctive telescope building, which resembles a huge white ski jump. Nearby are the domes of several large nighttime telescopes, all operated by the National Optical Astronomical Observatory.

Jack Harvey spent much of his early career working at the McMath-Pierce. He is perhaps best known for his research on solar magnetic fields, but he has many other interests, including helioseismology. Harvey is a talented experimentalist and a fierce competitor. Once he decides to tackle a problem, he becomes totally focused on it. He has built up an impressive record of groundbreaking research and was recently awarded the highest honor his solar colleagues can confer: the prestigious George Ellery Hale Medal.

In the early 1980s Harvey began to collaborate with Tom Duvall, a NASA scientist who was stationed in Tucson to operate a special solar telescope on Kitt Peak that maps solar magnetic fields. Duvall was originally interested in studying the large-scale flows at the surface of the Sun but soon became interested in the five-minute oscillations. As early as 1982 he displayed some of his keen physical insight. He showed how all the ridges in a diagnostic diagram (see fig. 3.5) collapse into one if the scales of the diagram are modified in a suitable fashion.

This transformation revealed a deep relationship among all the modes, which was dubbed "Duvall's law."

The two men became good friends. Harvey is tall, lean, and somewhat formal, while Duvall is shorter, plumper, and rather taciturn. Each has his own brand of humor, ironic in Jack's case, epigrammatic in Tom's. And each has a measure of experimental and analytic ability that makes them a formidable pair of researchers.

To obtain oscillation data they naturally turned to the large spectrograph at the McMath. They devised a clever scheme to determine how angular rotation speeds, averaged over all latitudes, vary along an equatorial radius. They formed an image of the Sun and positioned the slit of the spectrograph perpendicular to the Sun's rotation axis. A cylindrical lens was used to squeeze the entire disk into a line and to send it down the slit. In effect, they were selecting the so-called sectoral modes, which are north-south bands spaced around the equator (see fig. 3.3 with $L = 40$ and $M = 40$ for an example). They recorded the time variation of the Doppler shifts in two spectral lines, at two hundred positions on the east-west solar diameter, for eleven hours a day and for seventeen days. (Who said astronomy was an easy job?) When they were done they had a treasure trove of split frequencies for all degrees between $L = 1$ and $L = 100$. This broad range would allow them to probe a large part of the Sun, but without much resolution in depth and with no information about latitude variations.

Next, their theoretical colleagues stepped in with their heavy mathematical machinery. And what a collection of theorists it was, names we will meet over and over again, including Woitek Dziembowski, Philip Goode, Douglas Gough, and John Leibacher. These wizards collaborated in modeling the frequency data Duvall and Harvey had gathered (see note 5.2 for more details). Recall that there are two ways of analyzing such frequency observations: (1) construct a solar model, complete with assumed solar rotation and predict the expected oscillation frequencies, changing the assumptions if necessary; or (2) combine the modes in such a way as to isolate a narrow shell in the Sun, using a solar model as a guide, and then invert the observations to deduce the rotation speeds.

These theorists followed the first route. Figure 5.4 shows their result, the radial variation of rotation speed at the solar equator, as published in *Nature* in 1984. What do we see here? The angular speed is virtually constant at the surface

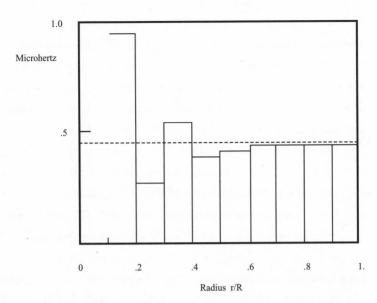

FIG. 5.4 A 1984 estimate of the depth-variation of rotation speed, at the Sun's equator. The vertical scale shows the rotation frequency, or reciprocal of the rotation period. The dotted line shows the rotation frequency at the surface of the Sun.

value, all the way down to a fractional radius of 0.4, and then it does a funny wiggle in the core. The core speed seems about twice as large as the speed at the surface, although it has a large uncertainty. These early results were enough to raise some doubts about the established picture of an angular rotation speed that decreases inward, as Glatzmaier and others had found from their simulations. The steep rise in the core was plausible because, as mentioned in chapter 4, the Birmingham group had seen it in their low-L data.

This was a good start toward a map of solar rotation, but as noted Duvall and Harvey could say nothing at this point about variations with latitude; that would remain for another bright young man to discover.

ANOTHER MOUNTAIN HEARD FROM

Timothy Brown has always had the good luck (or the good sense) to work with outstanding colleagues. First he was a student of Henry Hill and helped to re-measure the oblateness of the sun (see chapter 4). With Hill and another student, Robin Stebbins, he analyzed the oblateness data and found a number of long-period oscillations of the Sun's diameter. After he received his doctorate in 1976 he joined the Sacramento Peak Observatory in New Mexico, where he continued to work in helioseismology. There he encountered two other clever men, John Evans and Jacques Beckers, both experienced experimentalists.

Brown transferred to the High Altitude Observatory after a few years but returned to New Mexico to make observations. Beginning around 1980, he, Evans, and Beckers built a new type of detector (the Fourier tachometer; see note 5.3) to make two-dimensional velocity images of the solar disk. This instrument samples all points on the disk simultaneously, unlike a spectrograph, which builds up an image line by line. It is capable of measuring solar velocities smaller than a meter per second over long periods. Rugged and very stable, it was the prototype for an instrument aboard the Solar Heliosphere and Oscillation satellite (SOHO), launched in 1995.

In 1985, Brown reported the first results with this novel device. He had observed for five successive days and obtained the frequencies of all the modes with degree L between 8 and 50, as well as all their split frequencies. From these data he could immediately deduce that rotation in the radiative zone (between a radius of 0.3R and 0.7R) varies far less with latitude than at the surface. In other words, the differential rotation fades away in the deep interior. Moreover, the rotation period in this zone was close to the equatorial period at the surface: twenty-seven days. But Brown's data were still not good enough to say much about rotation in the convection zone above a radius of 0.7R.

There was more to come. But first Tom Duvall and Jack Harvey made a big splash.

BREAK OUT THE ANORAKS

It was obvious to everyone: to map the rotation of the sun's interior, you had to measure accurately the splitting of oscillation frequencies. The only way to do that was to obtain the longest possible continuous strings of data. Gerard Grec, Eric Fossat, and Martin Pomerantz had shown how, with luck, that was possible at the South Pole. So in 1981, Tom Duvall and Jack Harvey packed their warm clothing and headed south.

Unlike all previous oscillation experiments, they would measure oscillations of brightness rather than velocity. Because sound waves compress and heat the solar gas, they cause changes in its brightness. When many waves overlap, the result is an oscillation of brightness by about five parts in a thousand. In comparison, the velocity signal of many overlapping modes is much larger, about one part in three or four. Velocity measurements are therefore intrinsically easier— the signal is stronger. But because the oscillations are nearly vertical at the surface, the velocity signal fades out near the edges of the solar disk. Intensity oscillations don't do that, so a larger fraction of the disk becomes useful, allowing an easier separation of modes. In addition, a variety of effects can change the shape of a spectral line and introduce errors in a measurement of a Doppler shift.

Duvall and Harvey decided to observe the intensity oscillations in a small piece of the solar spectrum (about 1.2 nm wide centered on the calcium "K" line at 393 nm). They planned to record the oscillations with a solid-state detector, which had some 40,000 pixels to cover the solar disk, and to use the same tiny telescope that Pomerantz had built for the French expedition. They packaged all the rest of their electronics in one small crate that could survive the weather (fig. 5.5). Like the French, they set up their gear a considerable distance upwind from the diesel generator that supplied their power.

Life at the Amundsen-Scott station was hardy but bearable. Their quarters were about the size of a broom closet, and they had to get used to twenty-four hours of sunshine and learn to fall asleep without darkness. The meals were excellent, and everyone was fit and healthy, in part because germs can't survive the cold. The station had a marvelous mix of scientists from several disciplines, including biologists, geologists, and aeronomists. But there was little time to socialize. They worked fourteen-hour days, fought the equipment, and prayed for sunshine.

However, not everything was deadly serious at the station. Robin Stebbins, a colleague of mine from Sacramento Peak, had spent the previous austral summer at the pole, trying to get his oscillation experiment to work, and he told me about the "One Hundred Degree Club." On December 21, midsummer at the

FIG. 5.5   Duvall and Harvey posing with their equipment at the South Pole in 1981.
Duvall is on the left.

pole, the hardiest (and foolhardiest) of the men would soak themselves in the
sauna until they were bright red. Outside, in the frigid cold, stood a barber-
striped mast, the unofficial "South Pole." The game was to leap out of the sauna
and race naked once around the pole before one froze. Somehow, I can't imagine Harvey or Duvall participating in such frolic.

A six-day period of clear weather began on November 16, 1981. When the ice
fog and blowing snow were gone, the polar sky was an incredible deep blue, comparable to that in the Himalayas. The sun shone twenty-four hours a day, only
thirteen degrees above the perfectly level horizon. Unfortunately, Harvey and
Duvall arrived two days late and struggled with their equipment throughout the
rest of the clear period. They never experienced as long a stretch of clear weather
thereafter, but they did succeed in obtaining fifty hours of uninterrupted data,
the best third of their observations.

Their oscillation spectra showed tens of thousands of modes, a huge catch that presented an awesome task of reduction and identification. They had captured all the modes with degrees between 20 and 98, which would enable them to probe, for the first time, the rotation speeds throughout the convection zone at fractional radii between 0.7 and 1.0. Moreover, they could map the rotation in latitude.

A complete analysis of their data took four years, but they were able to report in 1986 that the differential rotation throughout the convection zone was *hardly different from that of the surface*. That was totally unexpected and in direct conflict with theoretical models. Gary Glatzmaier's simulations, for example, predicted that the rotation speeds in the convection zone are constant on *cylinders* (see fig. 5.3), while the South Pole results implied they are constant on *radii*.

So the situation at this point was as follows: the pattern of differential rotation seen at the surface seemed to persist through the depths of the convection zone. Below the convection zone (whose base lies at 0.7 radii) the differential rotation practically disappears. This was great progress, but the emerging picture was still pretty fuzzy. It would take much longer strings of data to resolve the finer details.

ANOTHER ENTRY IN THE RACE

By the mid-1980s, observers had tried almost every trick in the book to peer into the Sun's interior. They had used spectrographs, the Fourier tachometer, resonance cells, an oblateness detector, and a two-dimensional photometer.

Ed Rhodes was betting on yet another type of device. He began to collaborate with Alessandro Cacciani, from the University of Rome, who had developed the magneto-optical filter. Like the Birmingham instrument, it uses resonance scattering in a sodium vapor to detect Doppler velocities, but it produces a *two-dimensional* Doppler image, not just an average over the solar disk (note 5.4). Rhodes and his colleagues set up this filter at the eighteen-meter tower telescope at Mount Wilson Observatory. In 1984 they obtained observations over sixteen consecutive days, but it took them another three years to analyze the data. They had measured the frequency splitting for all degrees from L = 3 to L = 170, a huge task. Stephen Tomczyk, a graduate student of Roger Ulrich's at UCLA, used the same equipment for his thesis and obtained an independent set of frequency splittings.

Altogether, these data yielded few surprises but added confidence to the findings of Duvall and Harvey. The analysis indicated a slow inward decrease of the rotation speed in the convection zone and hinted that the surface pattern of differential rotation fades out the deeper one looks. We'll hear more from Rhodes and from Tomczyk in chapter 9.

TIMOTHY BROWN GETS HELP

Tim Brown was fortunate in convincing a bright graduate student, Cherilynn Morrow, to undertake her doctoral thesis with him as supervisor. Morrow was enrolled at the University of Colorado and somehow resisted all the suggestions from the great crowd of theorists there to do a theoretical rather than an experimental thesis. In the end she made a wise choice. Experimental theses often take longer than theoretical ones, because you have to get your equipment built and debugged. But by 1984 Tim Brown had succeeded in doing that and the rest was pure gravy.

They used the Fourier tachometer at Sacramento Peak and succeeded in getting a run of fifteen clear days in October 1984. They were able to resolve all the splitting frequencies for all degrees between 10 and 100. Then they compared their results with those predicted by Jørgen Christensen-Dalsgaard for various profiles of rotation.

They reached two main conclusions. First, the radiative zone (between 0.3R and 0.7R) rotates as a solid body, which was rather surprising in itself and raised a basic question: What frictional force ties together the different layers in this zone? Second, they determined that the convection zone (down to a radius of 0.7R) rotates differentially in latitude at angular speeds that resemble those of the surface. In that respect they agreed with Duvall and Harvey, and with Rhodes and Tomczyk. They cautioned, though, that it was too early to say that the surface pattern persists unchanged throughout the zone. Better data would be required to pin down the radial variation. And it was not long in coming.

BIG BEAR STRIKES AGAIN

The white dome of Big Bear Solar Observatory rises like a mirage from the middle of Big Bear Lake in the mountains east of Los Angeles. Harold Zirin, an unconventional professor at Caltech, picked this unconventional site in order to avoid the air turbulence on land that usually degrades the quality of solar images. He made an excellent choice. His observatory is famous for extended periods of fine viewing. Moreover, during the summer, Big Bear enjoys an unbroken succession of sunny days.

In 1983, Caltech hired Ken Libbrecht as a junior faculty member who would work with Zirin. Libbrecht had been a student of Robert Dicke and had participated in his program to measure the solar oblateness. Dicke was a superb experimentalist and he trained a select company of students to follow in his footsteps. Libbrecht was one of the best—smart, competent, and confident. When he arrived at Big Bear, he resolved to make a mark in helioseismology.

For many years, Zirin had used a type of narrow-band optical filter to make solar movies in the light of H alpha, a strong spectral line of hydrogen. This "birefringent" filter was invented in the 1930s by Bernard Lyot, a creative French astronomer, and independently by Ingve Öhman, a Swedish scientist. Libbrecht modified one of these filters to transmit a particular spectral line of calcium. Then he added an optical switch that allowed him to sample the red and blue wings of the line in succession. The difference of the light intensity in the two wings is a measure of the Doppler velocity of the gas that emitted the line (see fig. 4.6). With this arrangement and a digital camera, Libbrecht was able to observe oscillations over the whole disk of the Sun.

In addition to all his other virtues, Libbrecht is both industrious and lucky. He observed tirelessly every day from March 26 to August 2, 1986, and obtained one hundred days of useful data. Of course he had the usual nighttime gaps, but his 60,000 full-disk images guaranteed the highest frequency resolution and the lowest background noise achieved to that date.

Libbrecht needed three years to convert his raw observations into identified modes. When he was done he was able to resolve individual azimuthal modes for

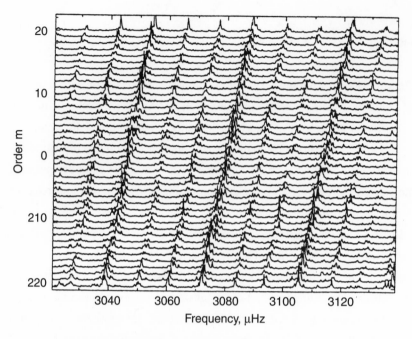

FIG. 5.6 Frequency splitting into modes because of solar rotation. Each mode is designated by three quantum numbers, N, L, M. The strong ridges correspond to modes with (N, L) = (15, 19), (15, 20), (15, 21).

the first time. Figure 5.6 shows an example of the splitting in frequency of the modes. Reading from left to right, the strong inclined ridges correspond to modes with (N, L) equal to (15, 19), (15, 20), and (15, 21). The weaker ridges between the strong ones are spurious and are introduced in the data by the day-night cycle. But the solar effects are clear.

If the Sun weren't rotating, these (N, L) modes would have the frequencies that correspond to M = 0, namely, 3047, 3080, and 3114 microhertz. The different M-modes along a ridge are separated in frequency by about 0.4 microhertz, which corresponds to a rotation period of twenty-nine days. So, somewhere in the sun lies a layer rotating at an average period of twenty-nine days. But where is that layer? And how do all the other layers rotate? It would take a full-scale inversion of Libbrecht's data (see note 5.2) to determine that, and to build a more detailed picture of the Sun's internal rotation.

Jørgen Christensen-Dalsgaard and Jesper Schou undertook that monumental task. In 1988 they displayed the new picture (fig. 5.7). As Tim Brown had forecast, each layer in the convection zone rotates slower at the poles than at the equator, but this differential behavior fades out the deeper you look. Below the base of the convection zone lies the radiative zone, which rotates as a solid body would. Below a fractional radius of 0.4, nothing was known.

This new picture of the internal solar rotation has enormously important consequences, which Peter Gilman, Cherilynn Morrow, and Edward DeLuca were quick to point out. The basic point is that the radiative zone *rubs against the base of the convection zone*, at a radius of about 0.7. As we shall see in chapter 10, theorists require a *shearing motion* deep in the sun to generate the magnetic fields that we see at the surface. Inside the convection zone, shear is relatively weak, because each layer rotates much like its neighbor. But at the base of the convection zone, there is sharp transition from differential to solid-body rotation, and the shear is strong. The ideal location for the "solar dynamo" that generates magnetic

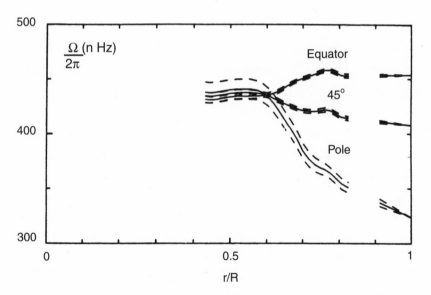

FIG. 5.7   Rotation periods in the Sun's convection zone, as derived from Libbrecht's 1986 observations. Note that the periods are constant along radii, not on cylindrical surfaces, as was formerly expected.

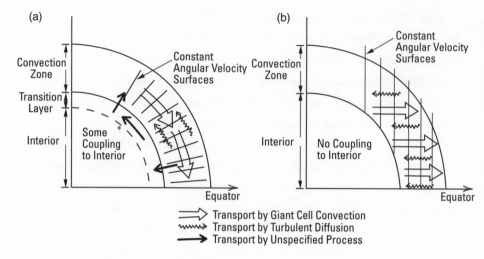

FIG. 5.8   The new (a) and old (b) pictures of how rotation speeds (or angular momenta) vary in the Sun's convection zone.

fields in an eleven-year cycle is at the base of the zone. This clue has been followed up vigorously, as we will see in chapter 9.

Then there is the whole question of just why the Sun chooses to rotate in this complicated fashion. Later on we'll look at recent simulations that try to answer that question, but in 1989 Gilman and company sketched a possible scenario, which figure 5.8 illustrates. They contrasted the new picture of constant rotation on cylindrical surfaces (5.8a) with the old one of radial surfaces (5.8b). They postulated that giant convection cells, which carry heat to the surface, also transport angular momentum from high to low latitudes. (How this is done would require a lengthy explanation.) The real mystery, for which nobody had a suggestion, is how the angular momentum is returned to high latitudes under the base of the convection zone (see the dark solid arrows) to complete a cycle. All this has become somewhat clearer in time, but still remains an open question.

AND WHAT ABOUT THE CORE?

In some ways the core of the Sun (inside a radius of about 0.3R) is the most important and least accessible part of the interior. It is important because only there is solar energy generated by thermonuclear conversion of hydrogen to helium. A beautiful theory had been constructed to describe this process, but although the theory accounted successfully for the Sun's luminosity, it predicted less than half of the observed flux of neutrinos (we will discuss all this in detail in chapter 7). Was the nuclear theory that far off? Or was something wrong with our ideas about the core of the Sun? It was a critical question for astrophysicists. Helio-seismologists were therefore eager to pin down the properties of the core.

How best to do that? The ideal way would be to observe the oscillations of gravity (that is, buoyancy) waves because they spend a lot of time in the core. However, they are so well confined that hardly any sign of them is expected at the surface. The next best approach is to observe sound waves (p-modes) of low degree (say L = 0, 1, 2, 3) because only these waves pass through the core. In chapter 7, we will see what we have learned about the core and how it bears on the missing neutrino problem.

The rotation of the core is also important, because it bears on the question of how an interstellar cloud redistributes its original angular momentum as it contracts into a star. Measuring the core's rotation is especially difficult because, as noted earlier, the frequency splitting patterns of the low L-modes are very narrow and require very long continuous data strings to resolve.

By the end of the 1980s the need for long, uninterrupted data strings became critical. Astronomers found ways to obtain them, as we will see in the following chapter.

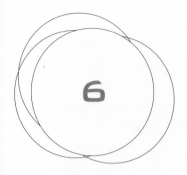

6   BANISHING THE NIGHT

BY THE MID-1980s, everyone recognized that future progress in helioseismology would depend on obtaining more precise oscillation frequencies. All the good things one wanted to learn about the solar interior (its rotation, composition, temperature profile, internal magnetic fields, for example) depended on resolving individual (N, L, M) oscillation modes. That meant that the splitting of frequencies had to be determined to within a gnat's eyebrow. And that, as we have seen, requires long, continuous observations, the longer the better, without any interruption from the setting of the Sun. In short, astronomers had to banish the night. And to do that, they would have to learn to organize and cooperate. A new era was arriving in which the lone researcher, using his own instrument, would become the exception.

Astronomers pursued three main routes to their goal. First, they ventured to the South Pole in the balmy austral summer, when the Sun skims above the horizon for six months. Second, they combined the observations from two or more well-separated locations—a network in which at least one site would always see daylight. And third, they launched their instruments into space, far from the nighttime shadow of Earth.

In this chapter we will follow their pursuit of long, long strings of data.

THE SOUTH POLE: PLUSES AND MINUSES

In principle, the Sun should be visible continuously for six months at the South Pole. In practice, a variety of factors limit the intervals of uninterrupted sunlight. High winds, snow squalls, cirrus clouds, and ice fog are the most serious obstacles. But even in clear weather, the low temperatures, high altitude, and difficulty of working in heavy clothing hamper the struggling astronomer. And if a crucial piece of equipment fails, there is little chance to repair or replace it in time.

We have already recounted the trials and tribulations of Gerard Grec and Eric Fossat, who had an unbroken run of five days. In their first attempt, Tom Duvall and Jack Harvey were less lucky and came away with only 50 hours (see fig. 5.5) at the pole. In 1987 they returned with Stuart Jefferies and obtained three runs of about 50 hours each over a period of 325 hours. Then in 1989, they won the brass ring: a run of 343 hours, with only 55 hours missing. Altogether, Duvall visited the pole five times but never exceeded this record. (As an award for his perseverance and his important scientific contributions, a mountain in Antarctica was named after him. That's almost as good as a Nobel Prize.)

Given the slim chances of exceeding even a five-day run and the difficulties of observing at the pole, only a few astronomers were willing to invest the time and energy. A recent example is David Rust, from the Johns Hopkins University. He conceived the Flare Genesis Experiment, a balloon-borne observatory that circled the South Pole in 19 days at an altitude of 38 km in order to observe the evolution of solar magnetic fields. But he is an exception. In the mid-1980s, most other astronomers were ready for another path to long-observing runs. They would create networks of ground-based observatories.

HANDS ACROSS THE SEA

The Birmingham group blazed the trail. They first set up identical resonance cells at Izana in Tenerife and the Pic-du-Midi in France and observed intermittently from 1976 to 1979. These two sites are separated by only one hour of longitude but have different weather patterns. Then in 1980, they moved from Izana to Mount Haleakala, Hawaii, a 3000-meter extinct volcano. Their sites were then nine hours apart, with excellent summertime weather. Once, they were able to observe up to twenty-one hours a day for eighty-eight days with this setup.

The Stanford Solar Observatory followed soon afterward by cooperating with the Crimean Astrophysical Observatory, ten hours of longitude away. Each observatory used a modified magnetograph to detect long-period solar oscillations, particularly the puzzling 160-minute period, for several years. Ed Rhodes and Roger Ulrich, whom we have met before, were observing oscillations with a magneto-optical filter at the sixty-foot tower at Mount Wilson Observatory. Impressed with the success of Stanford, they too formed a partnership with the Crimean Observatory. Their two-station net is called the High Degree Helioseismology Network, or HiDHN. Not to be outdone, Fossat and Grec of the Nice Observatory set up resonance cells at Izana and at the Pic-du-Midi observatory to improve their weather prospects.

These pioneers soon recognized, however, that to obtain complete twenty-four-hour coverage, more stations would be needed. The U.S. Air Force had learned that lesson long ago. It was interested in avoiding disruption of critical radio communications and damage to its satellites by solar flares. So in the late 1970s it built the Solar Optical Observing Network, with six stations distributed around the globe, to observe and report flares in real time. The facility cost millions of dollars and required a large staff of military personnel. Astronomers might dream about a similar network for helioseismology, but despite the public's growing interest and the high regard among physicists for the quality of science being produced, the prospects were dim.

Their networks could be developed only incrementally, as private funds or government grants became available. To attract funds they had to offer a special talent, technique, or idea. The Birmingham and Nice groups, for example, specialized in whole-Sun observations of low-degree oscillations; the Big Bear group in the interpretation of mode amplitudes and excitation. To reduce operating costs, these groups relied on the cooperation of host universities. For example, the Birmingham group enlisted the aid of Barry LaBonte and his students at the University of Hawaii to operate one of their resonance cells and return the data.

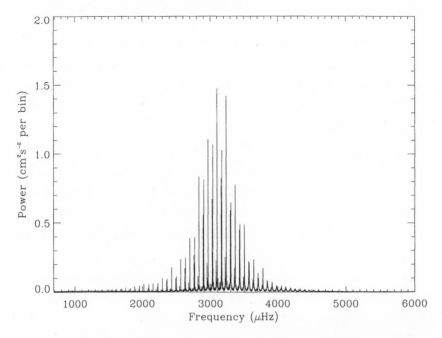

FIG. 6.1 Seven years of observations by the BISON network yielded this superb spectrum of the oscillations of the whole Sun. The background noise is minimal.

The French were among the first to build up a multistation network. Fossat and Grec organized an international consortium to build a six-station net called International Research on the Interior of the Sun (or IRIS). The stations are in Hawaii, Chile, Morocco, Spain, Uzbekistan, and Australia. The net started with two stations, which began making observations in 1984.

Around 1990, the universities of Birmingham and Sheffield joined to establish the Birmingham Solar Oscillations Network, or BISON. This is a six-station net with sites at Birmingham, Izana, South Africa, Australia, California, and Chile. Not all the instruments are identical, however. Figure 6.1 shows a superb oscillation spectrum of the whole Sun based on seven years of BISON observations.

Both IRIS and BISON employ resonance cells, observe the Sun as a star, and focus on low-L oscillations. In contrast, the Taiwan Oscillation Network (TON), planned for six sites, collects *resolved* images of the solar disk and measures high-

L oscillations of intensity of the calcium K line at 393.3 nm. This was the technique pioneered by Duvall and Harvey in their first trip to the South Pole. The first site, in the Canary Islands, was installed in 1993. Other sites followed in 1994 at Beijing, Big Bear in California, and Tashkent in Uzbekistan.

GONG

All these networks are busy churning out new data on oscillations. Each one has attracted a client group of analysts and theorists who wish to interpret the data. But only one six-station network operates all day, every day, for a full eleven-year solar cycle: the Global Oscillation Network Group, or GONG. (By now you have no doubt noticed the penchant of astronomers for catchy acronyms. If the particle physicists can invent terms like "charm" and "flavor" for charged particles, why shouldn't astronomers have some fun? Besides, a memorable acronym is helpful in dealing with funding agencies and the public.)

GONG was built by the U.S. National Solar Observatory (NSO) and has its headquarters in Tucson, Arizona. NSO, like the other national observatories, is funded by the U.S. National Science Foundation (NSF) to provide observing equipment for qualified scientists of any nationality to attack large-scale problems in astronomy. Understanding the interior structure of the Sun is one such problem.

GONG is the largest single group of helioseismologists, with over 130 participating scientists from twenty countries. With its latest hardware upgrade, it has been collecting as much as 100 *gigabytes* per site per week, and still manages to reduce and archive the data in a reasonable time. The group has contributed some important insights, which we will turn to in the chapters that follow. But how could they convince the tight-fisted funding agencies to part with large sums of money? GONG's story is typical of most "big science" projects.

From a Gleam in the Eye to "First Light"

The idea for an oscillation network arose in April 1983 at a meeting of the solar physicists of two observatories, the Kitt Peak National Observatory in Tucson and the Sacramento Peak Observatory in southern New Mexico. At the time, the two groups were joined as a loose confederation, which in two years would evolve into a separate organization, the National Solar Observatory.

About a dozen scientists met in the cold, white interior of the McMath-Pierce Solar Telescope on top of Kitt Peak, about ninety km from Tucson. This huge solar tower is unique in having a sloping light path rather than a vertical one. It contains the longest light path (136 meters) and the largest solar mirror in the world (1.5 meters). Much of the apparatus in the tower is built on a gargantuan scale, and some has to be moved about with overhead cranes. The tower provided a fitting setting for planning an elaborate network of telescopes.

The original group included Jack Harvey, Robin Stebbins, Tom Duvall (as a NASA scientist), and John Leibacher, all veteran helioseismologists who had made important contributions during the past decade. The group recognized that anything they built with public money would have to serve the whole community of solar astronomers, not just themselves. That was the guiding philosophy at all the national observatories.

The first task was to draw up a set of scientific goals. Where was the field of helioseismology headed? Which problems in the structure and dynamics of the solar interior were most pressing? What kinds of observations would be needed to solve them? A team was organized to write down answers to such questions in fine detail. A second team, led by Jack Harvey, began to think about hardware. Of all the possible oscillation detectors (spectrographs, optical filters, resonance cells, magneto-optical filters), which one was the most likely to meet the scientific goals and stand up to rugged field conditions?

Then there was the ticklish question of how many stations would be needed and how to space them in longitude around the globe. Frank Hill, a recent graduate from the University of Colorado, and Gordon Newkirk, the late director of the High Altitude Observatory, tackled that question. In 1985 they simulated the performance of a sample network, using the most reliable climate data they could find. (Fig. 6.2 illustrates the huge improvement in the quality of the data that a network provides.) They confirmed that at least six sites were needed to see the Sun 90% of the time. But where should one locate six sites to guarantee optimum performance? In addition to lots of sunshine, the sites would need power, road access, and some on-site assistance. Clearly,

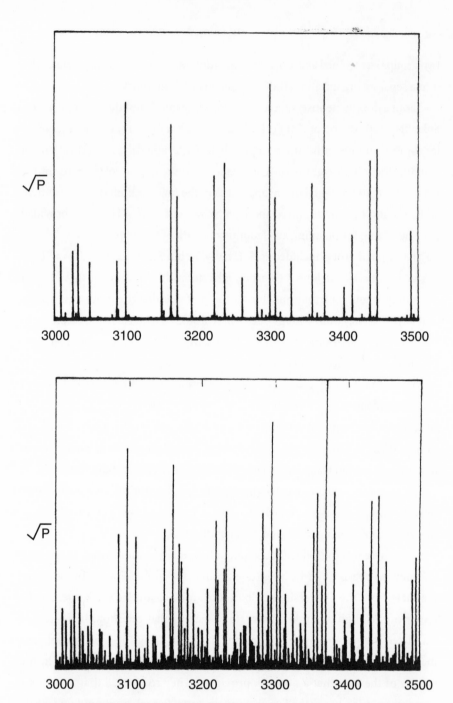

FIG. 6.2 The day-night interruptions of solar oscillations introduce a forest of spurious oscillation peaks. The bottom panel shows the spectrum that one year of observations from a single site might produce; the top panel shows the spectrum produced from an unbroken series of observations over one year. The comparison speaks for itself.

a list of candidate sites would have to be tested for several years to make an informed choice.

The amount of data such a network would produce was staggering. Early estimates topped a gigabyte a day. There was no point in turning on this fire hose without a plan to combine the images into a single uniform time sequence. Using techniques that Harvey and Duvall had designed, Hill assembled the calibration and reduction algorithms and eventually Jim Pintar, a crack programmer, wrote the software.

How much computer horsepower would be needed to reduce the data in a reasonable time? How would users access the data? The problem suggested that a central data processing center should be set up in Tucson, equipped with a battery of workstations. Users would download the data from a giant data bank.

Every aspect of the project bristled with technical and economic trade-offs. The staff decided to seek help. A Science Advisory Committee was organized to get unbiased opinions on a broad range of issues. Peter Gilman (High Altitude Observatory), Robert Noyes (Harvard University), Alan Title (Lockheed Palo Alto Research Laboratory), Juri Toomre (University of Colorado), and Roger Ulrich (University of California at Los Angeles) were invited to participate. Gilman, Toomre, and Ulrich are theorists; Title is an experimentalist; and Noyes does a lot of everything.

In the coming competition for funding, the National Science Foundation would want to know not only whether the project made good sense as science, but also whether astronomers gave it a high priority. The task of informing the community and marshalling support fell to John Leibacher. As a respected scientist, among the first to interpret the meaning of the oscillations, Leibacher had a lot of credibility. Moreover, and just as important, he speaks and writes well, thinks fast, and has a good sense of humor. At several large meetings of the American Astronomical Society he outlined what GONG could deliver and what it might cost. He claims his audiences needed no persuasion, but I suspect he's being modest. In any case, he got the grassroots support he needed.

By mid-1984, Leibacher had drafted a detailed proposal to the NSF, requesting funds for a start on a six-station network that would operate continuously for at least three years. He would act as the Principal Investigator, with overall responsibility for the project. Jack Harvey signed on as Instrument Scientist, Jim Kennedy was tagged as the Program Manager, and Frank Hill was Data Scientist.

At that time, the National Solar Observatory was one of three units in a super-observatory, the National Optical Astronomical Observatory (NOAO). Before a proposal could be forwarded to the Science Foundation, the GONG proposal would have to pass intense scrutiny by the NOAO board of directors. The scientists on the board were mainly nighttime astronomers or physicists, each with his or her own vision of the future.

A small cloud appeared on the horizon. Another group of astronomers within NOAO was proposing a large telescope, optimized for the infrared for the Southern Hemisphere.

GONG would eventually cost approximately $22 million to build and operate for three years. Projects of this size are funded relatively infrequently by the NSF, and only through a competitive process. Besides NOAO, NSF funds the National Radio Astronomical Observatory and the National Astronomy and Ionosphere Center. In 1985, each of these centers wanted a multimillion dollar project. The probability that NSF would fund more than one large project in a year was small. The probability that it would fund two *from the same organization* was practically zero. Ordinarily, the foundation would bounce back two proposals and demand a decision from the organization. NOAO's board of directors knew all this but was split over which big project to submit.

John Jefferies, the director of NOAO, was also faced with a difficult choice. As a solar physicist he was especially well qualified to appreciate the science that GONG could produce. But at the same time, the infrared telescope had great scientific promise and was supported by a very large group of nighttime astronomers. What to do?

Jefferies had had a lot of experience as a negotiator. As director of the Institute of Astronomy in Honolulu, he had converted Mauna Kea, the pristine extinct volcano on the island of Hawaii, into one of the world's great astronomical centers. Many of the huge nighttime telescopes there had been built by foreign nations. In the development of Mauna Kea, Jefferies had to deal with foreign astronomers, the Hawaiian legislature, the mayor of Hilo, and an assortment of environmental, historical, and ethnic organizations. Not only did he navigate through all the shoals, but in the negotiating process he also managed to secure observing time for his staff on each new telescope as the price of admission to the top of the mountain. Not an easy parlor trick! Faced with this Gordian knot, Jefferies submitted *both* proposals and challenged the NSF to fund them both because of their unusual scientific merit.

The NSF relies on a variety of sources for advice on which to base expensive decisions. Its long-term strategy is based on the recommendations of a blue ribbon committee of the National Academy of Science, which is convened *once a decade* to forecast where astronomy is headed and which tools astronomers will need. Unfortunately, the committee's last report did not mention a solar oscillations network, so the Astronomy Division of the NSF had to rely on its own advisory committee, which was composed of a dozen luminaries from academia.

Leibacher appeared several times before this august body to outline the science that GONG could produce and the NSO plan for construction. Patrick Osmer, a senior galactic astronomer from the NOAO, presented similar arguments for the infrared telescope. The committee was certainly aware of the other networks that were forming around this time and asked, reasonably enough, why the United States needed a national, as opposed to a private, network. Why were three years of operation necessary? Why so expensive a design? Were enough American astronomers interested in the subject to justify the effort?

The decision was suspended for almost two years. In April 1987, the officials at the NSF went behind closed doors and finally decided to fund the GONG first. Leibacher and his troops rolled up their sleeves and went to work.

The Instrument

Jack Harvey is arguably the best scientist at the National Solar Observatory. He can build instruments, observe with them, interpret the data, and argue convincingly with highbrow theorists. In 1985 he was the logical person to oversee the construction of the critical oscillation detector for GONG. But he had an unenviable decision to make: should he give up his own productive research for several years, in order to help a large group of potential users get the data that would advance helioseismology? In the end he did the altruistic thing, but it must have cost him sleepless nights.

FIG. 6.3   A pair of Michelson interferometer cubes, more precious than diamonds.

Of all the possible types of oscillation detectors, he rated the Fourier tachometer highest (see note 5.3). It was simple, rugged, stable, and sensitive. It produced resolved Doppler maps of the solar disk, containing the maximum amount of information. The lowest-degree oscillations (L = 0, 1, 2, etc.) might have to be sacrificed, but for intermediate and high L, the instrument was ideal. Tim Brown, Jacques Beckers, and Jack Evans had built a successful prototype at Sacramento Peak and used it to great effect.

The heart of the instrument is a Michelson interferometer, a fussy little cube of precision optics (fig. 6.3) that must be manufactured and assembled with excruciating attention to detail. All Harvey had to do was design a new one (with Jack Evans's help), order the optical parts, build six identical copies, test them, work out the bugs, and mate them with their controls and solid-state camera. No conceptual problems were involved, but the task occupied him for two years.

The basic strategy for the GONG instrument is complete automation. The oscillation detector and all its electronics are housed in a weather-proof metal shack that can be shipped as a cargo container (fig. 6.4). All it needs on the site

FIG. 6.4   Six GONG instrument shelters, lined up at a test site before deployment.

is electrical power and sunshine. The unit receives sunlight from an external alt-azimuth turret designed by Dick Dunn. The Doppler signals from the interferometer are read out with a solid-state camera (charge-coupled device), digitized, and stored on a magnetic tape. Once a week, someone from a local cooperating institution comes by to install a new tape and mail the old one to Tucson. All the heavy computer reductions are performed there on ganged workstations. But the flow of data doesn't end there. To serve the hundred users, GONG distributes data from an archive on demand, along with software and analysis tools.

Where to Put Them?

Frank Hill was given a nasty optimization problem for homework. Choose six sites from a list of fifteen that are distributed as uniformly as possible in longitude and latitude, that have the most sunshine all year round, and that lie within easy commuting distance of an airport. The network of six must see the sun at least 90% of the time, year in and year out. Hill loves computer games, but he admits this game challenged him.

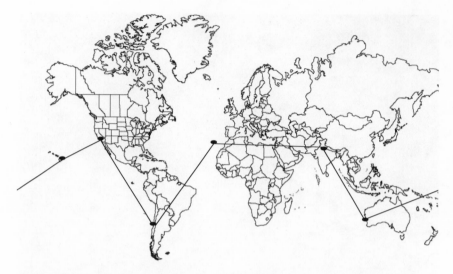

FIG. 6.5   The GONG network. Reading from left to right, the stations are located at High Altitude Observatory, Mauna Loa, Hawaii; Big Bear Solar Observatory, California; Cerro Telolo Observatory, Chile; Observatorio del Teide, Canary Islands; Udaipur Solar Observatory, India; and Learmonth Solar Observatory, Australia.

Cloud cover is relatively easy to find in this era of satellites, but reliable records of daily sunshine are not. So Tuck Stebbins and summer student George Fisher designed and built a collection of identical automatic "sunshine meters." These were one-inch telescopes that tracked the Sun throughout the day and recorded the brightness of sunlight as a voltage. A meter was set up at each candidate site and run all during the years that the hardware was built.

Frank Hill started with six possible sites and later added nine more in places like western China, Morocco, Saudi Arabia, and Hawaii. With all the necessary climate and insolation data in hand, he tested two hundred combinations of six stations. He ended up with the optimum network (fig. 6.5), which he predicted would see the sun 94% of every twenty-four hours, on average.

Note that each station, save one, was located at an existing astronomical observatory. This is no accident. These observatory sites were originally chosen with the same set of criteria that GONG had adopted. Furthermore, building a new site, possibly with a new road and utilities, would have been too costly, even for rich Americans. A seventh station was set aside in Tucson as an emergency spare.

The GONG instrument packages were built over a period of eight years, with an annual allotment of money trickling in from the NSF. Big Bear Solar Observatory received the first instrument in February 1995, and the others followed at intervals of a few months. By October 1995, the full system, including the data processing pipeline in Tucson, was complete and ready to go.

The network has met Hill's expectations, seeing the sun more than 90% of the time. Figure 6.6 shows the frequency resolution that combining three years of continuous data yields. You can easily see the splitting of the L-modes along each ridge in the diagnostic diagram. Later on we will discuss some of the science that such data have revealed.

After a first run of three years (1995–1998), NSF was persuaded to replace GONG's solid-state detectors with some that have four times the spatial resolution, for high-L observations. Now each site collects 100 gigabytes a day, a torrent that must be handled with additional computer power. Following the discovery that oscillation frequencies change during the solar cycle, and that the site of the solar dynamo had been isolated, the NSF agreed to continue to fund GONG into the future. The network is pumping out data even as you read this book, so the taxpayers are receiving a lot of new science for their money. You can follow the progress online at www.gong.noao.edu.

A SATELLITE OF ONE'S OWN

Ground-based networks have worked supremely well, and at relatively moderate cost. But no single network is able to observe all oscillation frequencies, all angular degrees (L's), all the time. A space observatory that carries several instruments above the turbulent atmosphere and out of the nighttime shadow of the Earth can, in principle, do just that.

But satellites have their own drawbacks. They are far more expensive than networks, their coverage of the Sun is limited by access to ground-based telemetry facilities, and if they fail in space, they cost a mint to fix. (Remember the repair of the Hubble Telescope?) Sometimes their instruments are several years behind the state of the art because they take so long to build and launch. Nev-

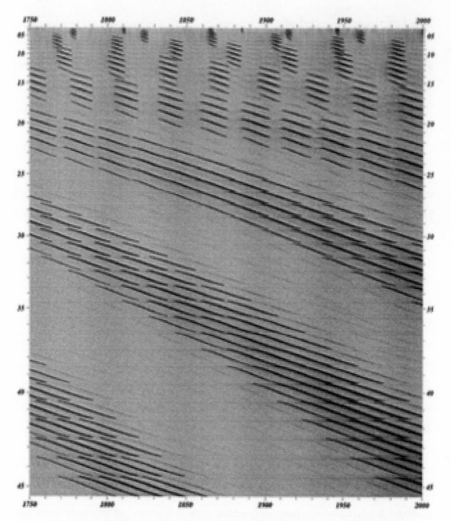

FIG. 6.6 A small segment of a ridge in a diagnostic diagram that was constructed from three years of GONG data. The degree L is plotted vertically, and the frequency (from 1750 to 2000 microhertz) is plotted horizontally. The splitting of L-modes into M-modes is easily seen. The darkness of a line indicates the strength of the mode.

ertheless, helioseismologists have hungered for a satellite of their own since the early 1980s.

They finally got a share of one in the Solar and Heliospheric Observatory (SOHO) (fig. 6.7). SOHO is an international project led and built by the European Space Agency. NASA launched it on December 2, 1995, and the craft now orbits about the point in space (the so-called Lagrange point) about a million miles toward the Sun, where the gravitational pulls of the Sun and Earth are balanced. From there, the 1850-kilogram satellite has an uninterrupted view of the Sun. It has enough rocket fuel onboard to maneuver it for twenty years, but no decision has been made on how long to continue to take data. SOHO carries twelve instruments. Of these, nine carry out coordinated studies of the corona and solar wind. The remaining three experiments were built to study solar oscillations. Each experiment has a Principal Investigator (PI), who is the leader and spokesperson for a large team of scientists, engineers, and technicians. He or she won a place on SOHO through a rigorous competition of ideas and designs. To get onboard, each PI must agree to share the reduced and calibrated data with anybody who asks for it. In practice, the data are generally available over the World Wide Web, but one has to know what to ask for. This policy of instant availability is a departure from the old practice of reserving the data for the experimental team alone for at least a year, so that it could be the first to publish scientific results.

These experiments are not cheap. If an instrument fails in orbit, it can seldom be resurrected, and thousands of man-years of work are then lost. So absolute reliability, with redundancy, must be built in. The quality controls are slow, exacting, and expensive. A typical price tag reads "ten million dollars." Let's have a look at these crown jewels.

Global Oscillations at Low Frequencies (GOLF) is essentially a resonance-scattering experiment, similar in concept to the potassium vapor cells favored by the Birmingham group for their BISON network (see details in note 4.2). It views the Sun as a star, with no spatial resolution over the solar disk. Its main purpose is to probe the solar core with long-period sound waves and, hopefully, gravity waves. GOLF was designed to record low-degree oscillations with periods as short as two minutes and as long as one hundred days.

FIG. 6.7 Testing the SOHO satellite before launch.

Alan Gabriel is the leader for this experiment. He has had an interesting career, first as a theoretical spectroscopist interested in plasma fusion, solar flares, and other exotica at the Rutherford Appleton Laboratories in England. His skills as a diagnostician of hot plasmas made him valuable in a series of space experiments. Later he migrated to France, through a series of distinguished laboratories, and finally to Orsay, where he leads a team of astrophysicists involved in space research. GOLF is his latest enterprise.

Variability of Solar Irradiance and Gravity Oscillations (VIRGO) is the responsibility of Charlotte Fröhlich of the Physical-Meteorological Observatory, Davos, Switzerland. VIRGO's scientific objectives are similar to GOLF's, namely to record long-period, low-degree oscillations in order to probe the solar core. It covers the range of periods from two minutes to a day, with the aim of detecting solar gravity modes. Like its ancestor, ACRIM, it measures the amount of sunlight reaching the Earth (the "irradiance") with extremely high precision. In addition, VIRGO measures variations of the solar spectrum. Tiny fluctuations in the received light (a few parts per million) reveal the global oscillations.

Solar Oscillations Investigation/Michelson Doppler Imager (SOI/MDI) uses the Michelson interferometer that was used first in the Fourier tachometer and then as the GONG instrument. It records the usual five-minute p-modes, with good spatial resolution across the solar disk. It is most useful in probing the solar convection zone with oscillations of intermediate degree, say L = 10 to 200. The instrument also produces magnetic maps of the solar photosphere, with moderate spatial resolution.

Phillip Scherrer, the leader of the MDI team, has spent most of his career at Stanford University. He arrived there in 1974 after getting his doctorate at the University of California at Berkeley, and stayed on as research associate at the Institute for Plasma Research. For three years he headed the Wilcox Solar Observatory, gaining experience in solar magnetism and oscillations. He was one of the first young physicists to enter the new field of helioseismology and has made several important contributions. Scherrer is an active member of GONG and is highly skilled in data analysis and interpretation. The MDI is his first major space experiment. His proposal for the MDI aboard SOHO won over a number of strong competitors, largely because of its excellent scientific program.

Death and Resurrection

On June 24, 1998, after two years of flawless performance, SOHO suddenly went silent. Ground controllers at NASA's Goddard Spaceflight Center aimed telemetry signals, ten times the normal power, at the wounded bird but couldn't get it to respond. NASA engineers worried openly that it could be a total loss. The scientists who had delicate instruments aboard were devastated. They had hoped to observe through the peak of the current solar cycle in 2003. Now they could only take comfort in the knowledge they had had a good run.

All through July 1998, the satellite tumbled freely, slowly chilling down as its solar panels failed to point at the Sun. The instruments on board were going into deep freeze, with the danger that delicate optics would crack in their metal frames. On July 27, the 300-foot Arecibo radio telescope bounced a radar signal off SOHO, which the Deep Space telemetry network detected. At last the craft was found. It was still in its orbit, 900,000 miles toward the Sun, but tumbling at about one revolution per minute.

Finally, on August 4, controllers heard hoarse gasps from the satellite. The patient was barely breathing, but alive. Gingerly, engineers sent commands to thaw the frozen hydrazine gas using battery power, so that the solar panels could be turned to face the Sun. Slowly, the satellite recovered and all instruments were restored. Two of the three gyroscopes had failed, however, which has restricted the ability of the satellite to point in a desired direction. The near disaster was blamed on faulty pre-programmed commands, and on an inappropriate reaction of the ground crew.

In October SOHO returned to full operation. Then, on November 28, the craft went into "safe mode," shut down but still talking. The last working gyro had failed. A software patch enabled the craft to point in a crude fashion, but it went into safe mode again in December. Engineers cured the problem with more software on December 10, and SOHO has behaved well ever since.

It had been a roller-coaster ride. The series of malfunctions revealed the fragility of a complex spacecraft in space, but also the ability of engineers on the ground to work around problems. Men and women of NASA, we salute you!

• • • • •

As is evident, helioseismology evolved into a group effort in the late 1980s and early 1990s. No longer was it possible for one or two individuals to make a breakthrough with a single instrument, in a campaign of a week or two. The science had matured to the point where only continuous observations during months or even years could yield fresh results. This requirement drove solar physicists to band together into huge consortiums, to build large facilities, and to share the data democratically. In this respect, helioseismology came to resemble elementary particle physics, where large teams of experimentalists obtain data on specialized machines to supply the whole community of physicists.

Until recently, there was one exception. Stephen Tomczyk, Roger Ulrich's student, joined the High Altitude Observatory (HAO) in Boulder, Colorado, where he built a better mousetrap. This instrument is called the LOWL, and, as its name implies, it is designed to measure the splitting of the lowest L-modes (L = 0, 1, 2, 3). The lowest modes have the narrowest patterns of frequency splitting, and therefore are the most challenging. They are especially useful for exploring the temperature and rotation of the solar core.

The LOWL is based on Alessandro Cacciani's magneto-optical filter (see note 5.4) and has the advantage over the Birmingham resonance cells of providing two-dimensional Doppler images of the Sun, and therefore all the intermediate modes (L = 0 to L = 100). LOWL was placed at the HAO's solar station at 3300 meters on Mauna Loa, an extinct volcano in Hawaii. It ran off and on for six years (1994–1999) and, as we shall see, helped to nail down the model of the solar core.

Then in December 1999, Tomczyk completed an improved LOWL and set it up at Izana in Tenerife, where BISON and IRIS are observing. So now HAO has its own network, called Experiment for Coordinated Helioseismic Observations, or ECHO.

In the coming chapters we will look at some of the exciting results that are spewing out of these new facilities.

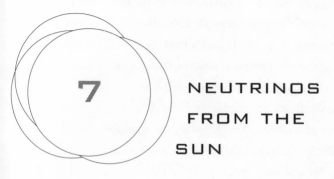

NEUTRINOS
FROM THE
SUN

AS YOU SIT READING THIS BOOK, 60 billion solar neutrinos pass unnoticed through every square centimeter of your body, every second. Physicists have told us these elementary particles have no mass, no electric charge, no magnetism, and travel at nearly the speed of light. They pass directly from the Sun's core, through the enormous cloak of gas that surrounds it, through the Earth and out to space, without ever colliding with anything. It would take a lead shield one light-year thick to stop one.

Wolfgang Pauli, one of the founders of quantum mechanics, was the first to guess that they existed. He conjured them up in 1930 to account for some missing energy as a proton loses a positron to become a neutron. He was later heard to lament, "I have done a terrible thing. I have postulated a particle that cannot be detected." Enrico Fermi, later famous for his role in developing the atomic bomb, named them "neutrinos."

Pauli was too pessimistic. In 1951, Fredrick Reines and Clyde Cowan, two physicists at the Los Alamos National Laboratory, realized that if nuclear theory was correct, the nuclear reactor at Savannah River, Georgia, should emit intense beams of anti-neutrinos. (Every elementary particle has a twin anti-particle. The positron, for example, is an electron with a positive electric charge. The anti-neutrino differs from the neutrino by the direction of a kind of spin.)

Near the reactor they set up a detector that contained fifty gallons of water. An anti-neutrino colliding with a proton in the water would release a pair of gamma rays. The gamma rays would produce flashes of light in a scintillating liquid, and a bank of photon detectors would trap the light.

This simple scheme worked beautifully. You may wonder why, since I just said a light-year of lead would be required to stop a neutrino, and the same applies to an anti-neutrino. But that is the worst case. Collisions follow the laws of probability. So although most anti-neutrinos pass through the target, a *very* few have a finite (but small) chance of stopping in a mere fifty gallons of water. Despite the enormous predicted flux from the reactor, only three or four collisions per hour were detected. Nevertheless, Reines and Cowan proved that anti-neutrinos (and therefore neutrinos) exist. Reines was awarded the Nobel Prize in 1995 for this work.

Later on, physicists learned that neutrinos come in three "flavors" that are associated with the three types of leptons: the electron, muon, and tauon. Although related, the three flavors of neutrinos don't behave in exactly the same way. And therein hangs a tale.

PREDICTING SOLAR NEUTRINOS

According to the modern theory of stellar structure, solar neutrinos are created in the complicated cycle of events in which four hydrogen nuclei (protons) are fused to form a helium nucleus (an alpha particle), with many intermediate products. (See note 7.1 for the gory details.) This cycle (the "proton-proton" cycle) was worked out in 1938 by Hans Bethe, the eminent Cornell physicist, and his colleague Charles Critchfield. Bethe received the Nobel prize for this work in 1967.

In the proton-proton cycle the whole is less than the sum of the parts. The helium nucleus has a slightly smaller mass than the four protons that were used to construct it. This "mass defect" appears in the form of energy. As Einstein wrote, $E = mc^2$. For every helium nucleus produced in the Sun's core, 26.7 MeV of energy and two electron neutrinos are created.

With the details of this cycle in hand, plus a lot of other physics (see note

7.2), one can build a toy Sun in a computer. The computer has to follow the evolution of the Sun from a cold uniform ball of interstellar gas, through a contraction phase, and into the long era of hydrogen burning. To start, we have to supply the original chemical composition of the gas. Then, at intervals of a few tens of millions of (simulated) years, the computer recalculates the changing temperature, density, and chemical composition distributions. The changes are slow enough to allow the Sun to be very close to equilibrium all the time. Usually, one assumes the Sun neither gains nor loses mass during its long adolescence.

Finally, at a simulated age of 4.6 billion years, we can compare the computed radius and luminosity of the Sun to their present values. If the predictions are wrong, we have to adjust the model, or add some new physics, and try again. We end up finally with a "model," a set of tables that describes every physical quantity at each point along a radius. Figure 3.6 shows the structure of the present Sun according to a 1988 model.

Astrophysicists have been playing this game for many years. In this way, they gave us their best estimate of what the interior of the Sun looks like. Mind you, this so-called standard model omits any mention of flows, of rotation or magnetic fields, and other important matters of interest to solar astronomers. Moreover, the details of the model continue to change as improved nuclear data or chemical composition are published, as better tables of the opacity of the gas to radiation are calculated, and so on. But the basic model was accepted as being close to the truth because it reproduced the two principle global quantities of luminosity and radius. That was about the best one could do, because as late as 1960 there was no way to check the model experimentally.

Then, in 1963, Ray Davis and John Bahcall, two young physicists at the Brookhaven National Laboratory, realized there was a way to trap an infinitesimal fraction of the Sun's huge predicted output of neutrinos. Since they knew the efficiency with which they could capture these few, they could extrapolate the number they caught to find the initial flux of neutrinos. And then they could compare their observations with the predictions of the standard model. At last it should be possible to find out if Bethe was right.

THE HOMESTAKE EXPERIMENT

Reines and Cowan had shown that, despite the anti-neutrino's elusiveness, it was possible to capture a few if one waited long enough with a sufficiently sensitive apparatus. Ray Davis thought he could do the same thing for neutrinos. His scheme depended on an exquisitely sensitive radiochemical technique. If a solar neutrino collides with an atom of chlorine, it will convert it to a radioactive form of argon, a "noble" gas. Its radioactivity tags the argon, like a pea in a bushel of rice, and Davis knew just how that argon atom might be extracted and counted.

Bahcall remembers how he and Davis were able to throw together their experiment without a lengthy proposal, extensive reviews by endless committees, environmental concerns, or special appropriations. The project was designed and built by a small team consisting of Davis, another physicist, a technician, and an engineer. The work was carried out within one and a half years, once the laboratory's director approved it and authorized a relatively small sum ($600,000) from the chemistry budget. (Contrast that with the campaign required to plan, fund, and build the GONG network!)

To catch even a few solar neutrinos would require a monstrous number of chlorine atoms. These were provided in a 100,000-gallon tank of a common cleaning fluid. To guard against false counts caused by cosmic rays, the tank was entombed at a depth of 1500 meters in an abandoned gold mine, the Homestake Mine in South Dakota (fig. 7.1).

After the equipment was set up, it had to be calibrated. A precise number of radioactive argon atoms were released into the tank and then recovered to test the efficiency of the radiochemical method. The radioactivity of the surrounding rocks also had to be taken into account. Finally, they were ready. They sat down to wait for a solar neutrino to stop in the tank.

Their experiment required the patience of Job, because the expected count rate was about thirty captures per year. No alarm bells go off when a capture occurs. They could only sit and wait and hope their equipment was still working. Then, after about two months of waiting, they had to find those three or four radioactive argon atoms in the 100,000 gallons of fluid. Just think how sensitive their method had to be!

FIG. 7.1   Chlorine neutrino experiment in the Homestake mine. The tank contains 600 tons of cleaning fluid.

The first results from the chlorine detector were announced in 1968. Davis expressed them in Solar Neutrino Units, or SNU (note 7.3). He had measured an average of 2 SNU during the past year. In contrast, Bahcall's solar model predicted 7.5 SNU, almost four times larger.

The astrophysical world was shaken to its core. How could the established theory of stellar evolution be wrong? Perhaps the experiment was flawed due to some systematic error? Could the nuclear rates that underlie the predicted count be that much in error? Or was the temperature model of the Sun wrong? That seemed unlikely, because the rate of energy production by the proton-proton chain varies as the fourth power of the temperature. Even a small error in the central temperature would lead to the wrong solar luminosity.

On the other hand, Bahcall had assumed that a second thermonuclear process, the carbon-nitrogen-oxygen cycle (also conceived by Hans Bethe), contributes only a small fraction of the Sun's energy production. This cycle was thought to matter only in stars much hotter than the Sun. Was it possible Bahcall had underestimated its contribution?

Davis went back to square one, checking his calibration and its reproducibility, worrying, agonizing; but in the end he was left with no choice but to believe his data. Similarly, Bahcall, collaborating with Roger Ulrich, tested the sensitivity of the standard model to various changes of input data and assumptions. Between them, they managed to reduce the discrepancy to a mere factor of two. At that point, the only recourse was to accumulate more data to reduce the random errors.

It seems a miracle that the experiment had worked at all, but it did so reliably for over two decades. Figure 7.2 shows the count rates during that time in SNU. The counts fluctuate, as you might expect from the low probability of capture, even dropping to zero at one point, but averaging about 2.5 SNU. For comparison, a 1978 version of the standard model predicted 7.56 SNU. The deficit of solar neutrinos stubbornly refused to go away. The "solar neutrino problem" took its place at the forefront of astrophysics.

Remember that in 1978, helioseismology was just getting off the ground. There was as yet no way to determine physical conditions in the Sun's core below a fractional radius of 0.3, where the sun's energy is produced. Only after the

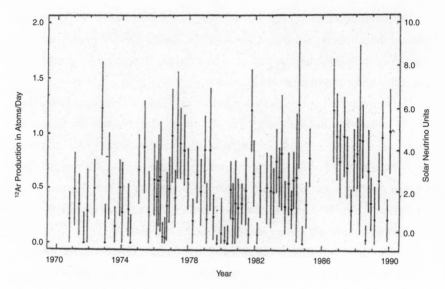

FIG. 7.2   Solar neutrino counts over twenty years.

multistation networks, the experiments aboard SOHO, and the LOWL had spewed out enough data would it be possible to get an empirical check on core conditions. Only when the predicted and measured oscillation frequencies agree to one-tenth of a percent could one have full confidence on the neutrino predictions of a solar model. Solar physicists would work for two decades to exceed that level of agreement between observed and predicted frequencies. At the same time, particle physicists pursued the elusive solar neutrino with ever more refined experiments.

In this chapter we follow both paths toward a surprising conclusion. Let's continue with the neutrino experiments.

NEW PARTNERS

Neutrino astronomy, as a new window on the universe, developed in parallel with helioseismology but far more slowly. Nobody but Ray Davis had an experiment running between 1968 and 1988. Although his annual counts fluctuated, the long-term average remained well below predictions. By 1985, particle physicists realized that the only way to crack the solar neutrino problem was to build new and different neutrino detectors.

Figure 7.3 illustrates the predicted energy spectrum of solar neutrinos. All of these are *electron* neutrinos, not tau or muon neutrinos. The great majority is produced in the first step of the proton-proton cycle, the fusion of two protons (labeled pp), at energies *below* 0.4 MeV. However, Davis's chlorine experiment detected only those neutrinos with energy *above* 0.8 MeV. Moreover, these neutrinos arise from two secondary steps in the proton-proton chain: the decay of an intermediate product (beryllium 8) and a proton fusion process ("pep") that involves an electron (note 7.1). A huge extrapolation was required to estimate the total flux of solar neutrinos from these neutrinos. In contrast, the direct fusion

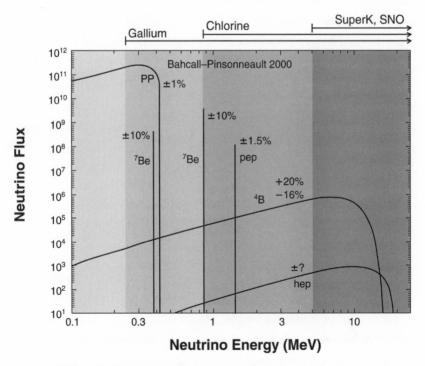

FIG. 7.3   The predicted energy spectrum of solar neutrinos.

of two protons by the pp process should produce ten thousand times as many neutrinos. They could yield a much more accurate estimate, but to detect them would require a different instrument that was sensitive below 0.3 MeV.

Bahcall and Davis argued that the best bet for that instrument was based on a reaction of a neutrino with an atom of the metal gallium. Gallium is rare in nature and therefore expensive. Moreover, a full-scale experiment would require three times the world's annual production of gallium. There was little chance these two physicists could corner the market, but they did manage to form a consortium and borrow 1.3 tons of gallium for a pilot program at Brookhaven. This group also developed the critical radiochemical technique that would be needed for the experiment.

The immediate prospects for a full-scale experiment in the United States floundered, however, in a quagmire of funding debates, departmental squabbles, and unending reviews. As Bahcall wrote, "Physicists strongly supported the experiment and said the money should come out of an astronomy budget; astronomers said it was great physics and it should be supported by the physicists" (*Public. Astrom. Soc. Pacific,* April 2000).

In the middle of this squabbling, the Russians seized the initiative. Ironically, with American collaboration, they built the Soviet-American Gallium Experiment (SAGE) in a deep mine in the Caucasus. A consortium of West European physicists followed soon after, and with American participation built GALLEX deep in the Apennines. Both groups reported their first results in 1992. Figure 7.4 indicates the energy ranges covered by these experiments.

THE JAPANESE ENTER THE FIELD

In my physics classes, we were taught that protons are forever. Left alone, undisturbed, they would always remain protons. However, physics moves on. One of the predictions of the Grand Unified Theory (GUT), which unites the strong, weak, and electromagnetic forces of nature, is the decay of protons.

A Japanese group determined to search for proton decay as a crucial test of the theory. Obviously, they needed to watch a large number of protons, so they buried a tank containing 700 tons of water in a zinc mine at Kamioka, 200 km west of Tokyo. Should a proton decay spontaneously into several lighter particles, these shards would create photons of blue Cerenkov radiation as they sped through the water, rather like the sonic boom a plane generates as it passes the sound barrier. An array of one thousand photomultiplier tubes was assembled to catch these photons and trigger an alarm.

The Japanese never saw a proton decay in six years of watching, but that very fact enabled them to set a lower limit on the proton's lifetime. The result was a staggering 10^{33} years (one followed by thirty-three zeros!). Our universe, by comparison, is a mere 10^{10} years old. One form of a GUT was disproved but others remain intact.

At that point, in the late 1980s, the Japanese decided to shift gears and attack the solar neutrino problem with their Cerenkov detection scheme. If a solar neutrino with an energy of at least 10 MeV were to bounce off an electron in a water molecule, the electron would recoil with enough energy to generate Cerenkov radiation, and the event could be detected. Moreover, the *direction* of the incident neutrino could be determined from the pattern of light.

Solar models predict that relatively few of these high-energy neutrinos leave the Sun's core (see fig. 7.3). Nevertheless, the Kamiokande experiment in Kamioka detected enough of these by 1992 to reach two important conclusions: the number is only 40% of the predicted number, and indeed they do come from the direction of the Sun. (It may seem surprising that anyone could doubt that, but there was always the chance that Davis could be detecting not solar neutrinos but radioactivity in the rocks around his tank.)

By 1992 the two gallium experiments had accumulated enough data to announce a result. Each reported counting only about half the predicted number of pp neutrinos (fig. 7.4). Since the pp neutrinos represent by far the great majority of all solar neutrinos, there was nowhere else in the energy spectrum to look for the missing ones. The solar neutrino problem was not going away.

RETURNING TO SQUARE ONE

Particle physicists began to wonder whether the problem lay not in the Sun, but in their established ideas of how neutrinos behave. As early as 1968, one year af-

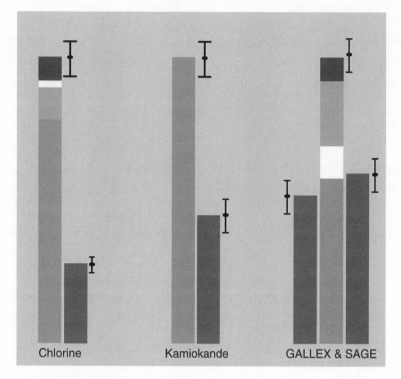

FIG. 7.4 Comparison of predicted (short bars) and measured (long bars) solar neutrino fluxes, from four different experiments.

ter the first results from Davis's chlorine experiment were published, Vladimir Gribov, a brilliant theorist, and Bruno Pontecorvo, a skilled experimentalist, suggested that some solar neutrinos might change identities after they leave the Sun.

As we mentioned earlier, neutrinos come in three "flavors" associated with electrons, muons, and tauons. The proton-proton cycle produces only the electron flavor. In the standard theory of the weak force, a neutrino preserves its flavor indefinitely once it is created. Pontecorvo and Gribov suggested something quite radical: solar neutrinos might *oscillate* from one flavor to another as they moved between the solar core and the Earth. If they happened to arrive at the Earth as tau or muon flavors, they would be undetectable by Davis's experiment. The idea was a bit too far-fetched to be considered seriously at the time, and in any case there was no way to test it experimentally.

Then in 1985, three Russians—Stanislav Mikheyev, Alexei Smirnov, and Lincoln Wolfenstein—predicted that this oscillation of flavors would be greatly enhanced should a stream of neutrinos pass through dense matter. This became known as the MSW theory. This mixing of flavors, though, could occur only if the flavors differ in mass. And as everybody knew, all flavors of neutrinos have zero mass. Or do they? Only a new experiment might tell. So particle physicists began to plan three large new neutrino detectors. They had at least three goals in mind: find the missing solar neutrinos if possible, test the MSW theory, and look for other cosmic sources of neutrinos, such as supernovas.

THREE NEW EXPERIMENTS

The four detectors described so far all suffer from the same limitation: they only detect *electron* neutrinos. That's a poor way to conduct a population census, because it ignores minorities. Moreover, they only deliver a single number (the total count) after weeks or months of counting. In contrast, the new detectors are able to capture *at least two* flavors of neutrinos. What's more, instead of ten events a month, the huge new machines collect hundreds. The direction and energy of each captured neutrino can be determined as well. These enhanced capabilities have allowed physicists to test their conventional theories of neutrino behavior.

The Super-Kamiokande, the pride of the Japanese, was the first and largest new detector to be built. It is a water-Cerenkov detector like its predecessor, but it is eighty times larger, containing 50,000 tons of purified water. This monster, forty meters in diameter and sixty meters tall, is built in two concentric layers. The outer layer acts as a shield for the inner one, and both are monitored by eleven thousand photomultipliers. The Super-K, as it is known in the trade, determines the flavor of a neutrino by the type of particle it produces in a collision with a proton. A muon neutrino produces only muons, an electron neutrino produces only electrons, and each of these generates a characteristic pattern of Cerenkov light. The pattern also reveals the direction from which the neutrino arrived.

The Canadian competitor is the Sudbury Neutrino Observatory. Like the Super-K, SNO is a Cerenkov detector, but it contains one thousand tons of *heavy* water, that is, water with deuterium atoms instead of hydrogen atoms. The Canadians had produced an ample supply of this expensive stuff for their reactor program. Heavy water gives the SNO the same collecting power as the Super-K, but with far less water. It is located 2 km down an old nickel mine in Ontario. The SNO is also impressive in size, with a diameter of twelve meters. Its ten thousand photomultipliers are mounted on a geodesic dome eighteen meters in diameter. SNO's advantage is a reaction with deuterium that detects *only* electron neutrinos. When salt is added to the heavy water, another reaction will soon be able to detect all three flavors of neutrino.

Borexino, the third of the new experiments, was designed for a special purpose. One step in the proton-proton cycle involves the capture of an electron by a beryllium-8 nucleus, with the release of an electron neutrino. The original four detectors measured far too few of these, compared to predictions, which was worrisome, because 15% of the Sun's energy output is supposed to originate in this step. Borexino was built to tie down the neutrino rate, and therefore the capture rate of this critical reaction. The hundred-ton device is located (you guessed it) underground, about 100 km from Rome, near the GALLEX detector in the Apennines.

Each detector is the brainchild of scores of scientists from a dozen or more countries and institutions. Only such cooperative efforts allow these complicated, expensive observatories to be built. Fortunately, they have uses for other projects besides the solar problem. For example, in 1987 the old Kamiokande detected neutrinos from a supernova in the Large Magellanic Cloud, emitted 170,000 years ago, confirming a current theory for the collapse of a stellar core, with the formation of a neutron star and a supernova.

All three of these new neutrino observatories would need several years to collect enough data to test whether neutrinos have mass and whether the MSW effect could explain the solar neutrino deficit. Meanwhile, the particle physicists had to be sure their predictions of the Sun's neutrino flux were reliable. They called for the best solar models that helioseismologists could provide.

PROBING THE SUN'S INTERIOR

As soon as solar physicists learned that the five-minute oscillations are standing sound waves, they realized how they could be used to probe the interior of the

Sun. One method was obvious. First, build a solar model in the usual way, so that it correctly predicts the Sun's present radius and luminosity. Then use the model to predict the oscillation frequencies of many different modes (note 7.4). If the predictions are off, change the model somehow. This is the so-called *forward* method of refining a model. It works, but it is laborious.

In 1976, only a year after Roger Ulrich's explanation for the oscillations was confirmed, two clever theorists proposed an alternative. Jørgen Christensen-Dalsgaard is a tall, cheerful researcher from Aarhus University in Denmark. Among the most prolific solar modelers, he pumps out three or four models a year as new data appear. Douglas Gough, a professor at Cambridge University, is his mentor and close collaborator. Gough is short, energetic, and brilliant, a fountain of new ideas. He is also a marvelous lecturer. When he delivered the prestigious Hale Prize lecture, he bounced across the stage swinging a pendulum to demonstrate a traveling wave. The audience of astronomers roared.

But, back to the story. These two theorists suggested that astronomers could borrow a tool from seismologists, the so-called inverse method. Instead of computing frequencies from a model of the Sun (the forward approach), *construct the model from the frequencies* (see note 5.2 for details). This technique of "inverting" the data is now used extensively and yields valuable quantities such as the sound and angular rotation speeds throughout the sun. These can be compared with the output of a model using the forward method.

The earliest inversions by Christensen-Dalsgaard and Gough demonstrated that the standard model was really quite good. It predicted the radial variation of the square of the sound speed to within one percent. (Recall that the square of the sound speed c^2 is directly related to the gas temperature.) As the precision of observations improved, though, that wasn't good enough. How could one change the model to improve the match?

In the early days, say 1982, a theorist could tweak the initial composition of the proto-Sun slightly, as this was fairly uncertain. The fraction of heavy metals, for example, could be shaded a bit. Eventually, Nicholas Grevesse and his Belgian associates closed this loophole with their meticulous solar spectroscopy. However, the initial fraction of helium remained a problem throughout the 1990s until the models themselves closed in on its value (0.275 by mass).

Modeling convection in the Sun is still a bit of an art. A semiquantitative theory exists, but it has a free parameter, the so-called mixing length. That is the distance, on average, that a rising convection cell travels before dumping its heat. The best guess is some multiple of the pressure scale height, the distance in which the gas pressure falls by a factor of $e = 2.72$. Each modeler is free to choose his or her own free parameter to improve a model's predictions, however.

The Sun's energy flows outward in the form of radiation, so the opacity of the solar gas at each wavelength must be calculated for each relevant combination of temperature and density. This is a huge task that has kept a small group of specialists slaving for over a decade. As improved atomic data are published, these indefatigable few revise their opacity calculations, then revise again. Nevertheless, at least three different sets of opacity tables are presently in use, and some give a better match with frequency observations than others.

Similarly, the equation of state—the relationship among temperature, pressure, and density—has been scrutinized. In the dense core of the Sun, a variety of physical effects come into play that had never been considered important before.

By 1990, the improved, scrutinized, and revised "standard" model was able to predict the sound speeds in the Sun to within 0.5% of those inferred from the observations. But the last small discrepancies were becoming harder and harder to beat down.

Then a new idea appeared: *diffusion.*

THE PERFECT CUP OF TEA

Drop a tea bag into a cup of hot water and you will see how the brown tea gradually spreads by the process of diffusion. Basically, the thermal motion of tea molecules allows them to wander in random directions. At any moment more tea molecules leave the bag than reenter it, and so the color spreads. This is molecular diffusion, and in the Sun it is too slow to separate one type of ion from another.

Back in 1958, Sydney Chapman showed another type of diffusion is much more effective in the Sun. Heavy ions tend to migrate toward higher temperatures, toward the core. So, for example, the helium ions in the well-mixed convection zone will diffuse downward into the radiative zone and then crowd into the core.

A few converts, like Arthur Cox and Joyce Ann Guzik at the Los Alamos National Laboratory, added another wrinkle, the gravitational settling of heavy ions.

Then, in 1992, John Bahcall and Marc Pinsonneault tried out these processes in a solar model. There was good news and bad news. Helium diffusion certainly improved the model's prediction of sound speeds. That was the good news. But unfortunately it also *increased* the neutrino fluxes, which were already too high. Charles Profitt, at NASA's Goddard Space Flight Center, agreed. Adding these effects raised the prediction for the chlorine experiment from 7.1 to 9.0 SNU, and for the gallium experiment from 127 to 137 SNU.

In a way, this was not entirely bad news. It was becoming increasingly clear that the solar neutrino problem was not caused by bad solar models but rather by incomplete neutrino theory.

TINKERING WITH THE SOLAR MODEL

By the mid-1990s the GONG, BISON, IRIS, HiDN, LOWL, and TON networks were all churning out oscillation frequencies. In addition, the three oscillation experiments aboard SOHO were performing beautifully. The journals were filled with new results, and topical meetings among hundreds of participants were being held several times a year. Helioseismology had become a cottage industry.

GONG presently produces 100 gigabytes of data per week at each of six sites. This flood keeps a small army of modelers, programmers, and technical assistants hopping, including a computer production line to grind up the raw data and spit out frequencies and mode identifications. The process is iterative; as the observations continue without interruption, month after month, the precision of the frequencies steadily improves. At intervals of several months, the modelers invert the data to extract a new profile of sound speeds and compare it to the predictions of their latest model.

By May 1996, the GONG network had been running flat out for seven months, and the team decided to show off their wares in a special series of articles in the journal *Science*. Figure 7.5 shows how changing the assumptions of a model changes its predictions of the sound speed. Each of the two panels shows the fractional difference between the observed and predicted speeds ([Sun − model]/

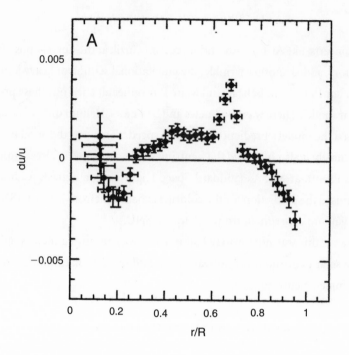

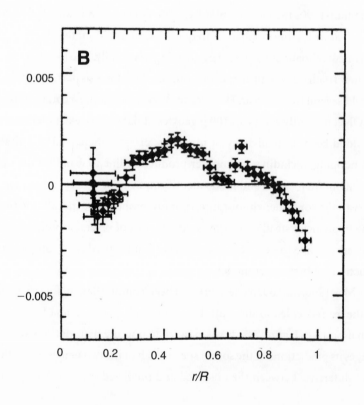

Sun). In this figure (as well as in figs. 7.6, 7.7, 7.8, and 7.9), perfect agreement at all positions along a radius would yield a horizontal straight line at a fractional difference of zero. As you can see, neither model in figure 7.5 achieves that desirable result, but neither deviates by more than about 0.4%.

Figure 7.5A tests the standard model with helium diffusion and gravitational settling included. It does surprisingly well, and the worst disagreement with observations occurs at the base of the convection zone at a fractional radius of 0.7. Something odd was going on there. Douglas Gough and his colleagues guessed that helium might be less abundant than the model assumes. Indeed, when they introduced weak mixing in this thin layer (fig. 7.5B) the bump disappears. They speculated that turbulence, induced by the known rotational shear in this layer, pumps helium from the radiative zone to the convection zone. The deviations in the Sun's core were harder to explain.

Not to be outdone, the MDI team published their sound-speed profile in 1997, based on two months of data. Figure 7.6 compares their result with a standard model. Alexander Kosovichev and his colleagues agreed with the GONG team that an excess of helium in the model could explain the big deviation at 0.7R. In addition, they suggested that an excess of helium at the edge of the core could account for the steep dip at 0.2R, and the steep rise toward the center could indicate that helium is *less* abundant than in the standard model. After four years of operation, GOLF and MDI data were combined to test a number of models that include turbulent mixing (fig. 7.7).

THE SPRINT TO THE FINISH LINE

During the past five years, a model of the Sun that satisfies everyone has finally been crafted.

Facing Page:
FIG. 7.5   The observed square of the sound speed, derived from the first six months of GONG data, compared with predictions from two models. The graphs show the difference, expressed as a fraction. (A) Model with helium diffusion and gravitational settling; (B) previous model, with mixing added. Sound speed is simply related to the temperature.

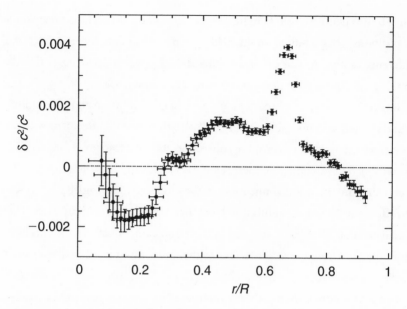

FIG. 7.6   A comparison similar to figure 7.5, with data from SOHO/MDI. The two experiments extracted virtually the same sound speed distribution.

Pinning down conditions in the core remained the most challenging problem. In 1996, Sarbani Basu and an international team at the Danish Center of Theoretical Astrophysics used LOWL frequencies to determine the sound speed in the core. They compared these with predictions from several models, looking for the golden combination of physical assumptions that would match the data. Their best model, including diffusion and settling, deviated by only 0.3% at 0.05R, deeper than anyone had probed before. They did even better in 1997 by combining LOWL and BISON data (fig. 7.8). As a further test, they predicted the neutrino fluxes that the chlorine and gallium experiments would measure, 8.5 and 13 SNU, respectively. These values, as we saw earlier, are still much too high. H. Antia and Shashikumar Chitre, among others, were coming to the same conclusion.

John Bahcall was ready to clinch the argument. He and his colleagues showed in 2000 that their latest solar model was finally accurate enough to predict reliable neutrino fluxes. Figure 7.9 compares the predicted and inferred sound speeds. The agreement is excellent. No curve deviates by more than 0.1% inside

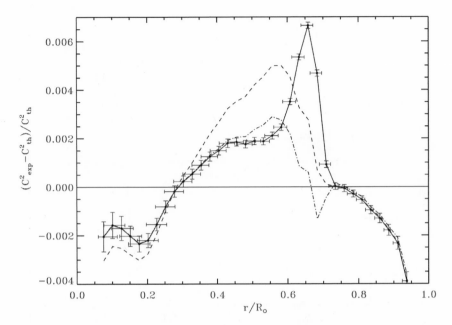

FIG. 7.7   Four years of MDI and GOLF data were used to extract the sound speed. The graph shows a comparison with models that include helium diffusion (solid line) and mixing at the base of the convection zone (dashed and dotted lines).

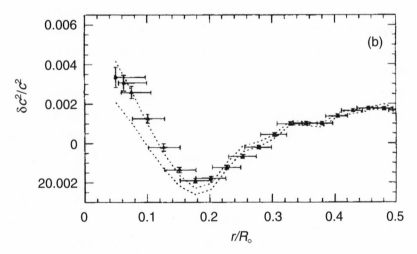

FIG. 7.8   The LOWL and BISON data yielded the best determination of the sound speed in the core.

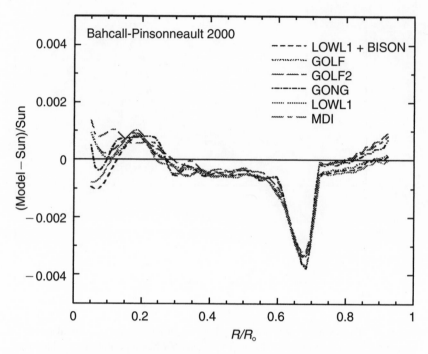

FIG. 7.9 One of the latest and most accurate models, compared with several sound speed inversions. The match is within 0.1% everywhere but at the base of the convection zone. The authors inverted the usual graph.

the critical range of 0.05R to 0.25R where neutrinos are produced. The predicted neutrino fluxes were still too high, however, so the answer to the neutrino had to come from neutrino physics.

Flavor oscillations and the MSW effect were the favored candidates for the new physics, but other ideas were floated. For example, if neutrinos possessed a magnetic moment, like little bar magnets, they might interact with the strong internal magnetic fields in the sun. Only new experiments could sort out the possibilities.

THE NEUTRINO SIDE OF THE HOUSE

Super-K began its search for neutrinos in April 1996. Within two years the team was ready to make an important announcement. They had discovered strong evidence that neutrinos do indeed *oscillate in flavor!*

Surprisingly, the data didn't involve the Sun, but cosmic rays (mostly energetic protons) instead. When a high-energy cosmic ray collides in the upper atmosphere of the Earth, it produces a shower of secondary particles. Among these are muon and electron neutrinos, in a predicted ratio of about two to one. Because neutrinos pass through the Earth so readily, they can arrive at the Super-K from a shower anywhere on Earth.

According to the Pontecorvo theory, however, a neutrino can oscillate among the three flavors. The probability that a neutrino will flip in flavor rises and falls periodically as it moves along its path to the detector. The distance between flips depends on several factors but is typically a few thousand kilometers. Neutrinos from a shower overhead travel only 15 km to reach the Super-K, but those originating somewhere on the other side of Earth have to travel as much as 13,000 km and have a better chance of flipping flavors. The team wondered whether they could detect a change of flavors by sorting neutrinos according to their incoming direction. Remember that the Super-K can determine a neutrino's direction from the pattern of Cerenkov light it produces.

Over a period of 535 days, the Super-K observed the direction of each cosmic ray neutrino as it arrived. Electron neutrinos arrived at the Super-K in equal numbers from every direction, but the number of muon neutrinos arriving from the far side of the Earth was only *half* of those from the near side. Evidently, muon neutrinos were changing flavors as they traveled the longer distances to the detector. Further tests indicated that the muon neutrinos were not flipping into electron neutrinos, but rather into tauon neutrinos, which the Super-K cannot detect.

In order for muon and tauon neutrinos to oscillate in flavor, they must have mass, contrary to the standard theory of the weak force. From their data, the team could only determine the *difference* of masses. They set lower and upper limits of 0.02 and 0.08 electron volts, or about a ten-millionth of the mass of an electron.

What about the electron neutrinos? Do they also oscillate in flavor? That was now the favored explanation for the notorious deficit of solar neutrinos. If that idea was correct, some of the electron neutrinos generated in the solar core have flipped into muon or tauon neutrinos by the time they arrive on Earth.

To find out, the Super-K team repeated the previous experiment but with a new twist. This time the team looked for a *day-night* difference in the arrival of solar electron neutrinos. The MSW theory, you will remember, predicts that flavor oscillations should be strongly enhanced as neutrinos pass through dense matter. During the day, the Super-K would detect a mixture of the electron and muon flavors as they arrived directly from the sun. During Super-K's night, on the other hand, the solar neutrinos would have to pass through the Earth to reach the Super-K and would have an enhanced chance of flipping flavors. If the MSW theory was correct, one should expect to see a *decrease* in the number of electron neutrinos during Super-K's night. For 504 days (1998–1999), the team looked for a day-night difference in the electron neutrino flux but couldn't find one. The experiment was not a total loss, nor did it disprove the MSW theory. Instead, it limited the range of certain critical parameters in the theory.

Finally we come to June 18, 2001, to what could be the solution of the solar neutrino problem. After two years of gathering data, the Sudbury Neutrino Observatory found direct evidence that solar neutrinos change flavors in their way to Earth. The full accounting of neutrinos has yet to be done, but it looks as though they have found the renegades.

How did they do it? Simple! SNO monitors a reaction with deuterium that occurs only if the incoming neutrino is an *electron* neutrino. The other two flavors can't trigger the reaction, so SNO yields an unambiguous count of electron neutrinos. On the other hand, the counts by the Super-K included an unknown fraction of muon and tauon neutrinos. So when the SNO team compared their count with that of the Super-K, they found that the Super-K's count was larger. And from the difference in counts, they were able to determine the fraction of muon and tauon neutrinos. It turns out that only a third of all the neutrinos arriving from the Sun are electron neutrinos. The remaining two-thirds have changed flavors and arrive as muon or tauon neutrinos. The missing neutrinos have been found!

In the laconic style of professional physicists, the 150-person SNO team wrote "The total flux of active ^{8}B neutrinos is thus determined to be $5.44 \pm 0.99 \times 10^6$ per square centimeter, per second, in close agreement with the predictions of solar models" (*Physics Review Letters*, June 2001).

In addition, the experiments estimate that the total mass of the three flavors of neutrino lies between 0.05 and 8.4 electron volts, or less than one-60,000th of the mass of an electron. That is not enough to account for the missing mass in the universe that cosmologists are looking for, but it could have an impact on the universe's expansion.

Bahcall was ecstatic. "I feel like dancing . . . For thirty-three years, people have called into question my calculations on the sun."

"The effect does not scream out at you from the data," Bahcall said. "You have to get down on all fours and claw through the details to see a small effect" (*New York Times*, June 19, 2001).

We can all cheer with John Bahcall. The prodigals have returned home.

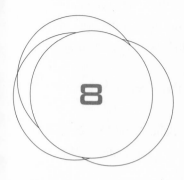

PICTURES IN SOUND

DOLPHINS DO IT, oceanographers do it, and physicians do it. Now he-lioseismologists do it, too. It's called "acoustic imaging" and it provides a way—actually three ways—to view isolated magnetic and thermal structures deep under the Sun's visible surface. In fact, it is even possible now to view things on the invisible *far side* of the Sun.

Everyone knows about sonar. In war movies we hear the ominous "ping" of the sonar pulse as it bounces off the steel hull of the submarine. The enemy destroyer overhead has located the sub's depth: it sent a pulse of sound down deep and timed the echo's return. But dolphins do it better. For years, biologists have known that dolphins use high-frequency squeaks to echolocate their prey, just like navy destroyers and bats. But Louis Herman, director of a dolphin research institute in Honolulu, has proved that the clever animals actually perceive an acoustic *image* in the echo. They are able to discriminate one complex shape (a Rube Goldberg contraption of twisted pipe, for example) from another, even in muddy water, purely from the returning signal. Nobody knows quite how they do it, yet.

Oceanographers also regularly use acoustic imaging to map the ocean floor. Their vessels steam along in a straight line and send sonar pulses sideways as they go. The echo delays give them details of the bottom hills and valleys in a broad band on each side of their track. And since the 1960s, physicians have used ultrasound imaging to view soft tissues, of which kidneys and pumping hearts are made, in the interior of the body.

In all these examples, the investigator (if we may include the dolphin) sends out a signal, times the return, and interprets the result. The shorter the wavelength of the sound waves, the higher the spatial resolution of the acoustic image. Hence dolphins use ultrasound, with frequencies ten times the upper limit of human hearing.

With no air between the Earth and Sun, there is no way a solar astronomer could probe the Sun's interior by sending a pulse of sound. But fortunately, that's not necessary. An astronomer can use the *Sun's own sound waves* to look for subsurface structures.

RINGS AND TRUMPETS

Way back in the early days of helioseismology, Edward Rhodes, Roger Ulrich, and Franz-Ludwig Deubner were the first to measure the rotation of the subsurface layers of the Sun (see chapter 4). They were able to do so because a sound wave acquires the motion of the gas in which it travels. Waves moving in the same direction as the Sun's rotation are speeded up, so an observer sees their frequency increase slightly. Conversely, a wave moving against the Sun's rotation is slowed down and its frequency appears to decrease. By comparing the frequencies of the two sets of waves, Rhodes and company were able to deduce the rotation speed within the layer in which the waves were traveling, and they were even able to assign a depth to that layer.

Frank Hill, the young scientist we met in connection with the birth of the GONG network, improved on this method. He was one of the first to use solar oscillations to look at small isolated structures embedded in the Sun, an example of "local" helioseismology. In the early 1980s theorists postulated the existence of giant convection cells, which extended the full depth of the convection zone. These cells were supposed to drive the differential rotation of the sun and also participate in the generation of solar magnetic fields. Hill set out to search for these cells. If he could find the horizontal gas flows associated with them, he could prove their existence. To search for such flows he planned to measure the frequency shifts of the sound waves within them, in analogy to the rotation study of Rhodes and his colleagues. A wave traveling with the flow would have its frequency increased, for example, and vice versa.

For three days Hill recorded the oscillations in a long rectangular strip along the solar equator. In his analysis he divided the strip into four square areas, each a quarter of a solar radius on each side. (This was the postulated size of one of these giant cells.) The usual procedure, which the Rhodes group had used, was to assume that the oscillation wavelengths are the same in the north-south direction as in the east-west direction. This procedure yielded a two-dimensional diagnostic diagram of oscillation power versus frequency and horizontal wavelength (see fig. 3.5). Hill dropped that assumption and created a *three-dimensional* diagnostic diagram.

Figure 8.1 shows a sketch of the surfaces of oscillatory power he found. They

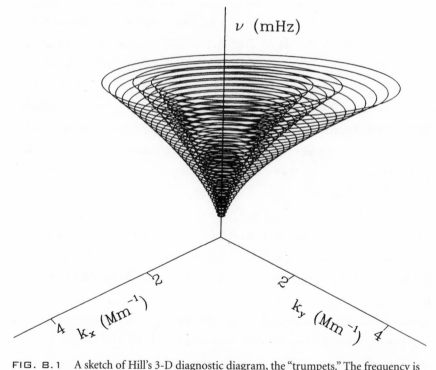

FIG. 8.1 A sketch of Hill's 3-D diagnostic diagram, the "trumpets." The frequency is labeled ν. K_x and K_y equal the reciprocals of the wavelengths in the x and y directions. So $K_x = 1/($wavelength in megameters, Mm$)$ for example.

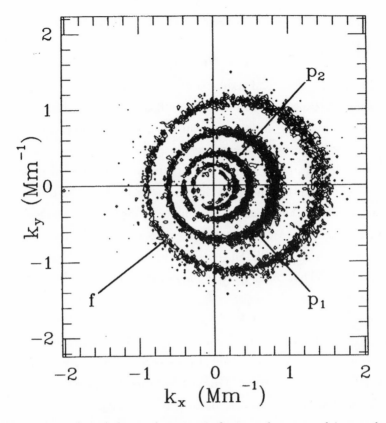

FIG. 8.2   A cut through the set of trumpets in figure 8.1 shows a set of rings, analogous to the ridges in a conventional 2-D diagnostic diagram. The displacement of the rings indicates a subsurface flow's speed and direction.

have the shapes of nested trumpets. Wave frequency is plotted vertically along the long central axis, and the horizontal wavelengths in two perpendicular directions are plotted on the two other axes. If we cut the trumpet at some point on the frequency axis, we find a set of concentric rings, as in figure 8.2. Each ring corresponds to a ridge in a conventional diagnostic diagram (see fig. 3.5). Note that the rings in this case are all shifted systematically away from the center of the diagram, which is an indication of both the speed *and direction* of a systematic flow under the surface. Hill's ring analysis has this advantage of yielding direction as well as speed. The depth of the flow can also be determined from the specific frequency of the set of rings. In this case, the speed was about 100 m/s, and the direction was roughly northwest.

Such speeds, however, were much too large to accord with the expectations of theorists. Indeed, Hill found his observations were plagued with tiny systematic effects and his results were discounted. Nevertheless, he had broken new ground by using sound waves to investigate a relatively small object embedded in the Sun. True acoustic imaging would come later.

In 1995, Jesús Patrón, a young Spanish astronomer, applied Hill's method to a new set of data and obtained a very interesting result. He and his colleagues (Ed Rhodes, Frank Hill, and Sylvain Korzennik) used the Cacciani magneto-optical filter (see note 5.4) at Mount Wilson Observatory to observe the oscillations in nine rectangular areas on the Sun over a period of three days. They determined the direction and horizontal speed of the gas, as a function of depth, in each of the nine areas.

Figure 8.3 shows their result. The gas in each area executed a *slow downward spiral*, like a whirlpool. The three vortices in the Northern Hemisphere rotated clockwise as they sank to 10,000 km. The six vortices on the solar equator or in the Southern Hemisphere rotated counterclockwise. This pattern suggested that the Coriolis force was responsible for the twist (note 8.1).

Patrón and company offered several interpretations of their result. The one I find the most plausible is spiraling convection. Numerical simulations of convection in a rotating fluid reveal fast plumes of gas that twist as they sink downward. It's possible that Patrón and friends observed a similar effect in the Sun. Giant convection cells, like those that motivated Hill, have never been detected, however.

SUNSPOTS UNDERCOVER

For many years solar astronomers wondered what a sunspot looks like under the solar surface. The magnetic field of a spot is concentrated in the dark center, the umbra. For a long while, the conventional model for the umbra under the surface was a solidly packed flux tube, like the trunk of a tree. In 1979, Eugene Parker of the University of Chicago, the dean of solar theorists, pointed out a problem

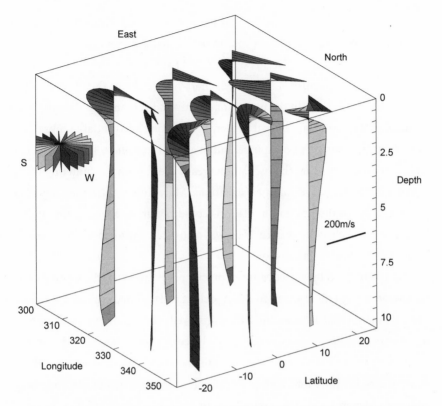

FIG. 8.3   José Patrón and collaborators derived this subsurface horizontal flow pattern from oscillation observations.

with this monolithic model. The strong magnetic field in the umbra would suppress convection, which normally requires circular motions to transport heat to the surface. The umbra would receive very little heat from below and would appear much darker than observations show. So Parker proposed a "cluster" or "spaghetti" model.

   In this scheme, the surface of the umbra would appear uniformly filled with a magnetic field, but just under the surface the umbral field would split up into a cluster of thin flux tubes, like the dangling arms of a jelly fish, with hot gas between them. These open spaces would allow some heat to reach the surface, in agreement with observations.

At the time, though, there was no way to confirm or reject either of these models. Then in 1982, Jack Thomas, a professor of mechanical engineering at the University of Rochester; Alan Nye, his student; and Larry Cram, a bright young Australian, had a brainstorm. Like many others before them, they had detected five-minute oscillations in the umbra of a sunspot. Everyone else thought these oscillations were spurious, caused by stray light leaking into the dark umbra from the bright surrounding photosphere. But in a paper in *Nature*, Thomas and collaborators suggested that the oscillations really arose from the buffeting of the sunspot by the Sun's five-minute oscillations. More importantly, they suggested that the study of the umbral oscillations could reveal the shape of the umbral magnetic field below the surface. Here was a possible way to decide between the two current models of a sunspot.

Thomas realized they needed more precise observations of the umbral oscillations. So he and his student Toufik Abdelatif collaborated with Bruce Lites at the Sacramento Peak Observatory. They observed oscillations in several umbras, using a spectral line that only appears at the cooler temperature of the umbra. From their two-dimensional spectral scans they were able to determine that a sunspot acts as a filter for incident sound waves, transmitting mostly long horizontal wavelengths and reflecting others. In addition, they found that the oscillatory power in the umbra was three to four times weaker than in the surrounding photosphere.

Abdelatif and Thomas were able to explain both of these effects, at least qualitatively, by modeling the umbra as a fully filled slab or cylinder of magnetic field. They calculated how the transmission of sound waves into the sunspot depends on frequency and how this behavior accounts for the observed deficit of power. They learned that only those sound waves with speeds greater than the characteristic speed of magneto-acoustic waves (note 8.2) inside the spot (about 25 km/s) are transmitted, while the rest are reflected. The transmitted sound waves are converted to magneto-acoustic waves inside the umbra and then stream up the umbra's magnetic field.

Barry LaBonte and Douglas Braun of the University of Hawaii picked up on this idea of using the global five-minute oscillations to explore the interior of a sunspot. With Tom Duvall, they tried out a new method. First they measured the five-minute oscillations in the vicinity of a sunspot. Then they divided the observed pattern of standing waves into two sets of traveling waves, with circular wave fronts. (This is always possible.) One set of circular waves converged radially on the sunspot from the outside, the other expanded radially from the spot. When they compared the acoustic power entering and leaving the sunspot, they found, as had Thomas and company, that a spot absorbs at least *half* of all the acoustic power incident on it. As we shall see, Duvall developed this technique of dividing incoming from outgoing waves into a very general tool.

WHAT'S REALLY HAPPENING UNDER A SUNSPOT?

Jack Thomas and friends were able to make a convincing case for the monolithic model of a sunspot. Their simple interpretation of their observations certainly went a long way toward validating the conventional view of a sunspot. In 1988 they were aware of Parker's alternative cluster model but they didn't try to test it. To do that required a detailed knowledge of how sound waves interact with thin tubes of magnetic field. That task was taken on by Tom Bogdan, a clever young theorist at the High Altitude Observatory. It would tantalize him for more than a decade.

Bogdan began by studying how a vertical slab of randomly spaced, thin magnetic flux tubes would scatter sound waves, and how it would compare to a monolithic slab, completely filled with magnetic flux. He learned that a dense cluster of tubes behaves very much like a fully filled slab, while a more open cluster has quite different transmission properties that depend on the spacing of the tubes and the wavelength of the incident sound waves. There was no simple explanation for the results of Thomas and colleagues.

Bogdan moved on in 1994 to consider more complicated interactions of sound waves and clusters of magnetic tubes. With two Belgian plasma physicists, Rony Keppens and Marcel Goossens, he studied the effects at the thin boundary layer between a magnetic tube and the surrounding field-free gas. They learned that strong "resonance absorption" of sound wave energy can occur in such thin layers. Absorption in this case means conversion of the sound waves into magneto-acoustic waves that ride up the bundle of flux tubes. A cluster model, it turned out, absorbs much more energy than a comparable monolithic model. The extra absorption occurs as sound waves are scattered repeatedly in the forest of flux tubes within a cluster.

All this argued well for the cluster model of an umbra. The problem, which remains to this day, is that even the enhanced absorption by a cluster of flux tubes is too weak to agree with observations. Bogdan eventually suggested that additional absorption occurs in the spreading magnetic field *above* the umbra. But it now appears that the absorption properties of a sunspot are not sufficient to allow a choice between the monolithic and the cluster models. A theory of convection in a strong magnetic field seems to be required, and we are not there yet.

Meanwhile, throughout the 1990s, another technique was evolving for studying isolated structures under the Sun's surface.

CAT-SCANNING THE SUN

Tom Duvall keeps hitting home runs. We have met him several times already— in connection with South Pole observations, in the founding of GONG, and with his famous law that explains why all the ridges in a diagnostic diagram fold into one. In 1993, he came up with another bright idea. It was similar to a method used by seismologists to probe the interior of the Earth.

When an earthquake occurs somewhere near the surface, it generates pressure and shear waves that propagate through the Earth. Some waves pass through the core, others through the mantle of the Earth. Seismographic stations spread around the world record the arrival times and strengths of these waves. By comparing the arrival times of waves with slightly different paths, seismologists can determine some of the physical characteristics of the different layers, such as their speed of sound, their temperature, and their chemical composition.

Duvall realized that a similar technique could be applied to the Sun. Instead of waiting for a rare "sunquake," set off by a solar flare, he could use the five-minute oscillations that are present all the time. As we have learned, sound waves are generated continuously by turbulent flows in the convection zone and some are trapped in resonant cavities in the sun. A typical trapped wave loops between the surface and the bottom of its cavity (see fig. 3.8). At the surface the steep drop

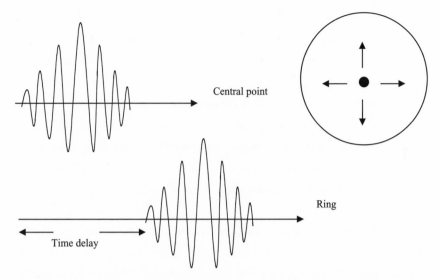

FIG. 8.4 Oscillations at a chosen point (top) are followed, after a time delay, by similar oscillations on a ring around the point (bottom), whose radius equals the horizontal wavelength of the traveling wave.

in gas density reflects the wave, and near the bottom the increasing temperature refracts the wave back toward the surface.

Suppose, thought Duvall, that the wave encounters a hot blob somewhere along its path. The speed of sound would be higher inside the blob than outside in the surrounding convection zone. As a result, the wave would return to the surface *sooner* than a similar wave that missed the blob. By timing the arrival of waves, one might be able to reveal hot or cool spots under the surface. Similarly, the gas flows inside a convection cell could accelerate a passing sound wave and cause it to arrive sooner or later. In this way, subsurface gas flows and possibly magnetic fields might be detectable.

The clever part of Duvall's method is how he finds the *same* wave at successive bounces at the surface. We can understand how he does this, at least qualitatively, if we imagine dropping a pebble into a pond. Circular ripples spread out from the center as the central splash continues to oscillate up and down. If we recorded the oscillations at the center and also at a point some distance from the center we'd find two curves (fig. 8.4) that look alike but are displaced in time. The displacement is obviously the travel time from the center to the external point. The farther the external point is from the center, the longer is the travel time.

In the Sun, a typical sound wave doesn't simply travel across the surface but loops beneath it. To find where a wave from some reference point returns to the surface, Duvall proposed to measure the oscillations on a set of nested rings centered at that point. By comparing ("cross-correlating") the oscillations at the reference point with those on each ring, Duvall could determine the particular ring in which the wave resurfaces, as well as the travel time to that ring. All the rings between the center and that special ring also oscillate, but without a good correlation to the central point.

In a 1993 paper in *Nature,* Duvall, Stuart Jefferies, Jack Harvey, and Martin Pomerantz reported how they put this simple idea into practice. They used 4096 full-disk images they obtained at the South Pole in 1988. Figure 8.5, taken from that paper, shows the result. Here we see the footprints of the same wave as it bounces back to the surface four times, at increasing distances from a chosen reference point. The arrival time at each bounce is carefully noted.

I remember hearing Tom Duvall present his method for the first time to a group of colleagues. He played it very cool, very understated. In a quiet monotone he outlined his result in a few paragraphs, showed a slide or two, and sat down. He left most of us with no idea of the potential power of his method. Perhaps he was unsure how far he could push it.

Three years later, Duvall and collaborators applied this "time-distance seismology" to plumb the regions under sunspots. They used 1017 oscillation images acquired at the South Pole in January 1991, a period when many sunspots were visible on the solar disk. In their analysis, they compared the oscillations *at each point* in the image with the oscillations on a ring of a chosen diameter that was centered on the point. For each point, therefore, they could determine the time a wave took to reach that ring and also the time it took a wave from the ring to reach the point. So, for each ring size they obtained two maps of travel time.

Now the radius of a ring corresponds to a particular horizontal wavelength. The larger the horizontal wavelength, the deeper a wave penetrates the Sun, as you can see from the sketch in figure 3.8. So as they increased the radii of the rings they were in effect selecting waves that returned to the surface from greater and

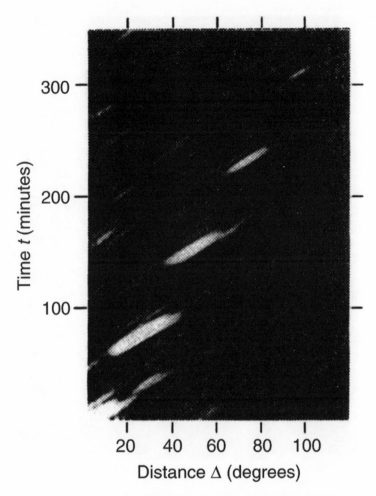

FIG. 8.5 A time-distance diagram, showing four bounces of a sound wave as it loops below the surface.

greater depths. Each ring yielded a different time-of-arrival map, as shown in figure 8.6.

What we see here are maps of the *mean* travel time (left column) and maps of the travel time *difference* between outgoing and incoming waves (right column). Remember that sound waves are "advected" or carried along by the gas in which they propagate. If a wave goes with the flow, it accelerates, while if it bucks the trend it slows down. By comparing the arrival times of incoming and outgoing

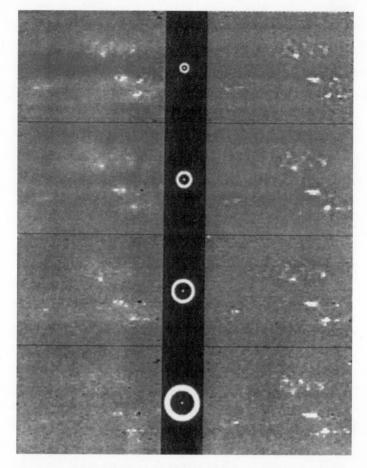

FIG. 8.6 Downflows under magnetically active regions around sunspots, detected by the time-distance method. Each panel in a column refers to a particular depth. The depths increase from top to bottom in a column. Bright regions in the time-difference maps (right column) indicate flows outward and downward from the active regions.

waves, Duvall and company were able to distinguish true gas flows from hot or cool spots in the gas, which affect incoming and outgoing waves in the same way.

In addition, the maps corresponding to different ring diameters refer to different depths under the surface. In this way Duvall and friends could map the flows as a function of depth. They had invented a method ("helioseismic tomography") similar in principle to the famous CAT scans that physicians use to create three-dimensional images of the human brain from a sequence of slices.

Bright regions in the time-difference maps (right column in fig. 8.6) indicate flows outward and downward from the magnetically active regions around the sunspots. Duvall's collaborators estimated downflow speeds of about one km/s, extending to a depth of a few thousand kilometers.

This type of tomography has become a powerful tool for probing the sub-surface structure of sunspots and active regions. For example, when SOHO/MDI oscillation data became available, Duvall and friends analyzed several days' worth of data to determine the temperature variations and flow directions down to a depth of 5000 km in the convection zone. Figure 8.7 illustrates their first result, obtained in 1997. Cool regions flow down in narrow funnels and hot regions spread horizontally. This result confirmed numerical simulations of convection performed by Bob Stein and Åke Nordlund, among others.

Later, Alexander Kosovichev from Stanford University, with a number of col-laborators, applied the method to more MDI data for a variety of objects. In

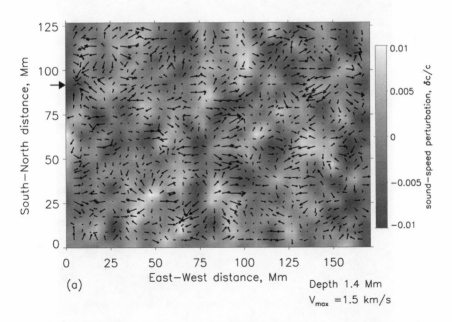

FIG. 8.7 The gas flows in the convection zone as revealed by the time-distance analy-sis of SOHO/MDI data. Distance is measured in megameters (million meters, Mm). The arrows indicate flow directions and the gray-scale indicates temperature.

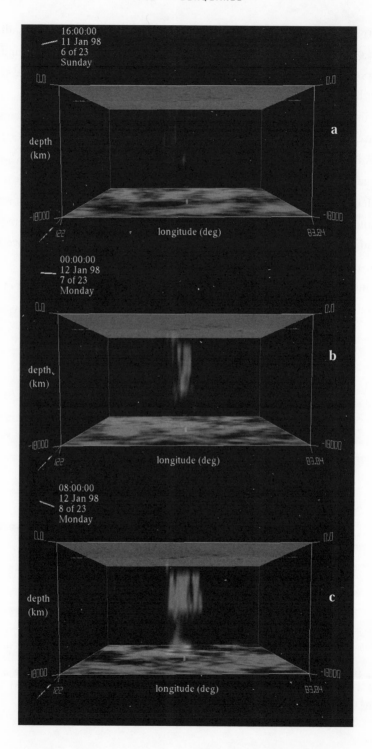

figure 8.8 we see the three-dimensional variations of sound speed (related to temperature) under an emerging magnetic region, down to a depth of 18,000 km, over a period of eight hours. At any moment, the sound speed varied from point to point by about 1 km/s. The whole region of strong magnetic flux, sunspots and all, rose at an average speed of 1.3 km/s and passed through the top 10,000 km in only two hours. That was much faster than theory had predicted.

Kosovichev and company also determined the sound speed variations under a sunspot, down to a depth of 24,000 km (fig. 8.9). The variations lie in the range of 0.3 to 1.0 km/s. The researchers estimated that a 1 km/s variation corresponds *either* to a 10% variation in temperature (2800 K at a depth of 4000 km) *or* to a magnetic field of 18,000 gauss. At the time they reported this research, they were unable to distinguish between these two possible causes of the variations.

In principle, the time-distance method should be able to decide which sound speed anomalies are due to subsurface magnetic fields. Sydney D'Silva and Tom Duvall have investigated how sound waves would interact with a strong magnetic field. They first predicted the time-distance curves for a quiet Sun, without any strong fields. Then they calculated the effect of a sunspot. Near a spot, they predicted that curves split into a family of closely spaced curves. The spacing and shape of these curves is sensitive to the shape of the field below the surface and could reveal whether the sunspot magnetic field is monolithic or clustered. However, as far as I can tell, nobody has managed to apply these predictions to real data.

Sunspots are not the only magnetic mystery that astronomers would like to explore. There is also the tempting target of the stronger magnetic fields far below the surface that finally emerge as sunspots. As early as 1983, Phil Goode of the New Jersey Institute of Technology and Woitek Dziembowski of the University of Warsaw looked into the possibility of detecting magnetic fields deep inside the Sun by observing changes in sound wave frequencies. They concluded

Facing page:
FIG. 8.8   Three-dimensional differences of sound speed under an active region that is rising from a depth of 22,000 km. Panel (a) shows the start, panels (b) and (c) are eight and sixteen hours later. In each box, depth (kilometers) is plotted vertically, and longitude (degrees) is plotted horizontally. The sound speed differences (shown as bright vertical strands) could be caused either by subsurface magnetic fields or by temperature variations.

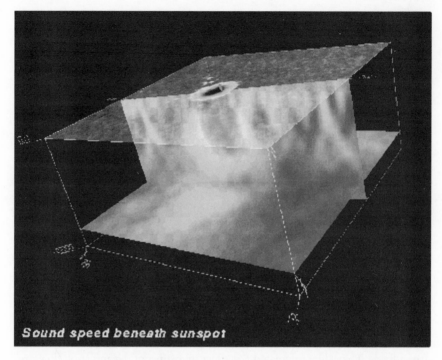

Sound speed beneath sunspot

FIG. 8.9 The 3-D sound speed variations under a sunspot, down to a depth of 24,000 km. Hot and cool regions are colored red and blue, respectively (see color gallery).

that the fields would have to have field strengths of millions of gauss to be detectable, a rather remote possibility.

The problem is complicated. Not only do strong magnetic fields scatter incident sound waves, they also oscillate. And if they do, they can produce a bewildering variety of magnetic waves and patterns. So, for the moment, direct detection of subsurface magnetic fields is still a distant goal.

3-D ACOUSTIC IMAGING

If solar tomography works, why not solar *holography?* That is the question Charles Lindsey and Doug Braun asked themselves sometime in 1989. These two clever scientists met while working at the University of Hawaii and have maintained a close collaboration ever since.

Most of us have seen a demonstration of holography, perhaps in a science museum (fig. 8.10). First, a laser beam is split into two beams. One part is directed to a solid object, where its rays are reflected and scattered in all directions. Some of these scattered rays fall on a photographic film. The second or reference beam is sent directly toward the film, without touching the object. The two beams *interfere* in the space between the object and film, producing a pattern that encodes all the spatial information of the object. The film records this interference pattern.

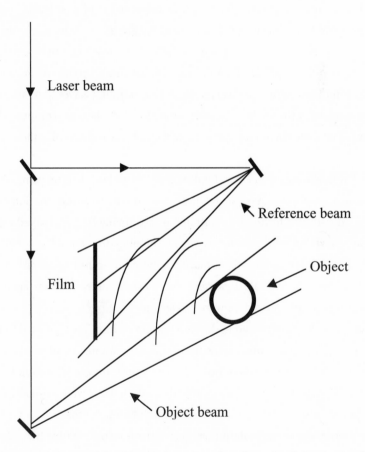

FIG. 8.10 The principle of laser holography. A laser beam is split in two. One beam illuminates the target, the other shines directly on photographic film. The direct and scattered laser light interferes at the film and encodes the spatial structure of the target.

To re-create the object, laser light is shined through the photograph toward the observer. The photograph acts as a diffraction grating, spreading the laser light in the same directions that the original incident beam took. The light rays overlap in space and reconstruct what appears to the eye as the original 3-D object. In fact, you can walk around the apparent object to view all of its sides.

The basic idea here is that if you can reverse the light you receive from an object, you can reconstruct the object in all its detail. For this to work in a museum, the light has to be *coherent,* that is, monochromatic and perfectly periodic in phase. The photographs (or holograms) taken during the experiment are then interference patterns that encode the spatial information of the object.

Lindsey and Braun recognized that something like a laser reference beam isn't necessary to make acoustic holograms of the Sun, because one can observe both the intensity *and* the phase of sound waves at the Sun's surface. As with time-distance tomography, one can use the Sun's own standing waves to image inhomogeneities under the surface. In effect, one has all the Fourier components of the waves that have encountered the subsurface object and can play them backward, like a movie in reverse.

In a landmark paper in 1990, Lindsey and Braun laid out the principles of acoustic holography and suggested that sunspots on the far side of the Sun might be detected with this technique. As Braun and company had discovered earlier, sunspots absorb half of the acoustic power impinging on them. That meant that sunspots would appear dark in an image composed of sound waves rather than light waves. And that tell-tale darkening could reveal a sunspot on the far side of the Sun, if the proper procedure were carried out.

Lindsey and Braun first had to develop that procedure. They decided to walk rather than run by performing some numerical simulations. What would a subsurface absorber of sound waves look like in a picture composed of sound waves rather than light waves? Figure 8.11 shows their result. Here we see two absorbers (the abbreviations SOHO and MDI) buried at different depths. A steady stream of five-minute sound waves illuminates them from below. By observing the oscillation pattern at the surface and calculating the paths the waves had to take to arrive at the surface, these researchers were able to reconstruct the absorbing objects below the surface. The absorbers appear as dark shadows. As the team

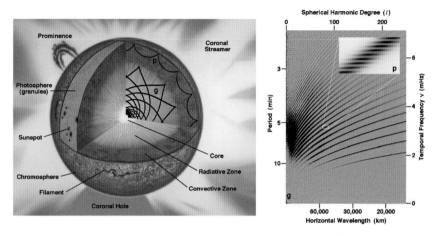

FIG. 2.2 A cutaway view of the Sun, showing its core, radiative zone, convective zone, and photosphere.

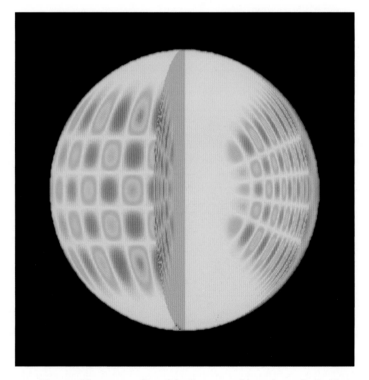

FIG. 3.4 If we could remove a slice of the Sun we could see the nodes inside it, spaced along a solar radius. In reality each interior node is a spherical surface.

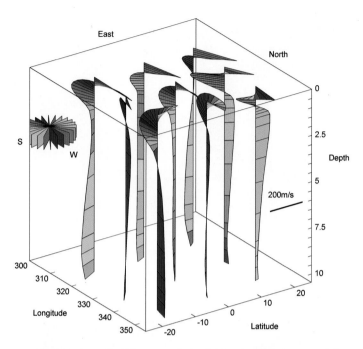

FIG. 8.3   José Patrón and collaborators derived this subsurface horizontal flow pattern from oscillation observations.

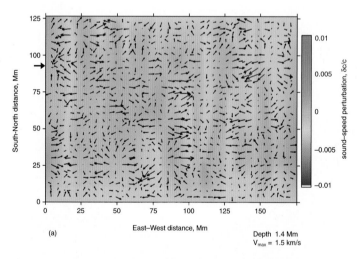

FIG. 8.7   The gas flows in the convection zone as revealed by the time-distance analysis of SOHO/MDI data. Distance is measured in megameters (million meters, Mm). The arrows indicate flow directions and the color scale indicates temperature.

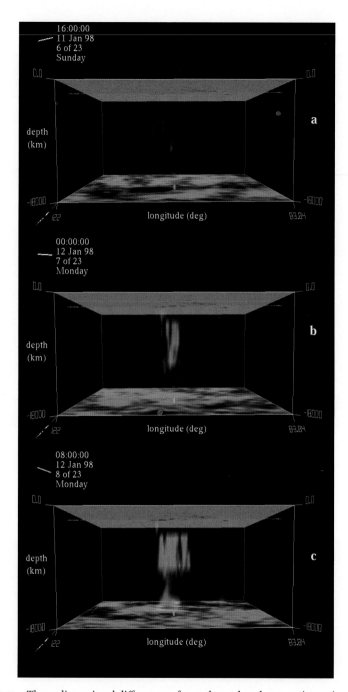

FIG. 8.8 Three-dimensional differences of sound speed under an active region that is rising from a depth of 22,000 km. Panel (a) shows the start, panels (b) and (c) are eight and sixteen hours later. In each box, depth (kilometers) is plotted vertically, and longitude (degrees) is plotted horizontally. The sound speed differences (shown as bright vertical strands) could be caused either by subsurface magnetic fields or by temperature variations.

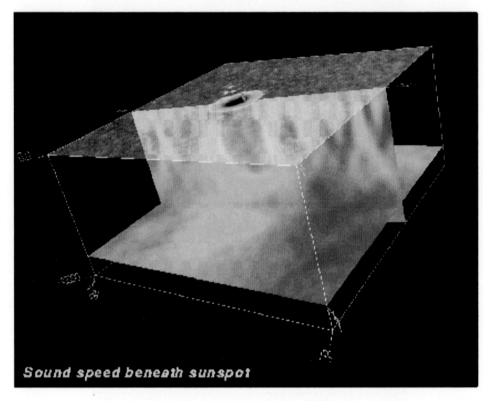

Sound speed beneath sunspot

FIG. 8.9   The 3-D sound speed variations under a sunspot, down to a depth of 24,000 km. Hot and cool regions are colored red and blue, respectively.

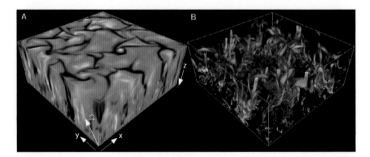

FIG. 9.10   In panel A we see a 3-D snapshot of the convection in a box (black areas are cool and downflowing). Turbulent downflowing vortices in panel B tend to align with the local direction of rotation, Ω.

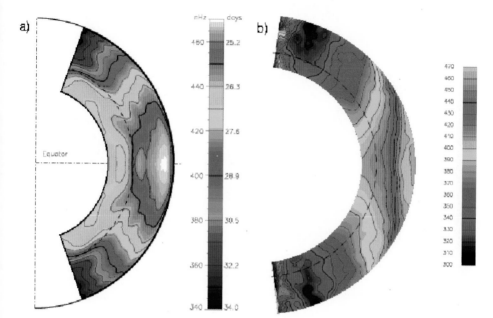

FIG. 9.12 A comparison of observed (a) and simulated (b) patterns of differential rotation. The angular speeds are indicated with a color code, red corresponding to rapid rotation and blue to slower rotation. The match is quite good, except for the remaining cylindrical pattern (the parallel vertical contours) in the simulation.

FIG. 9.13   A snapshot of turbulent convection in a box, computed at a *billion* points. This 1999 simulation at the Lawrence Livermore National Laboratory had ten times the spatial resolution attained at that time by solar physicists. The image shows tangling vortices and vortex sheets at one instant.

FIG. 10.1   These beautiful loops are formed by the magnetic field in the Sun's corona. The loops are filled with plasma at a temperature of about a million kelvin.

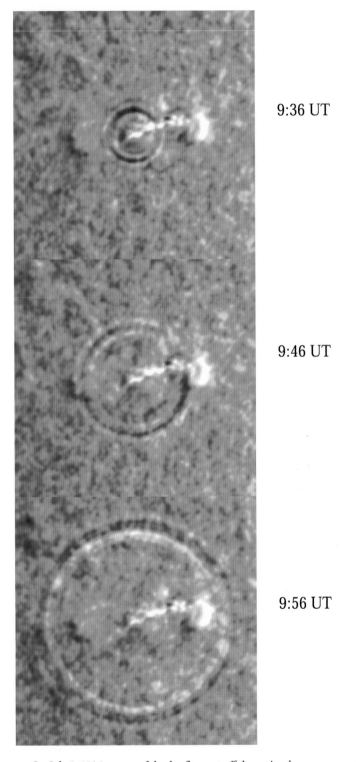

9:36 UT

9:46 UT

9:56 UT

FIG. 12.1   On July 9, 1996, a powerful solar flare set off these circular waves over the solar surface.

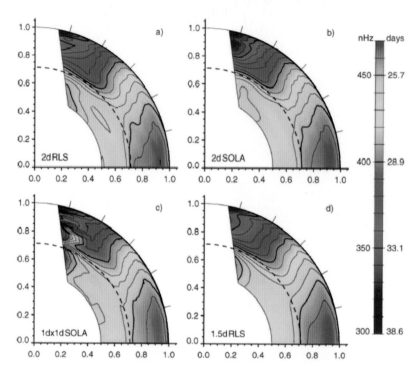

FIG. 12.5 The interior rotation of the Sun, determined by MDI data. Four different methods of analysis yield very similar results and set limits on the precision of the map.

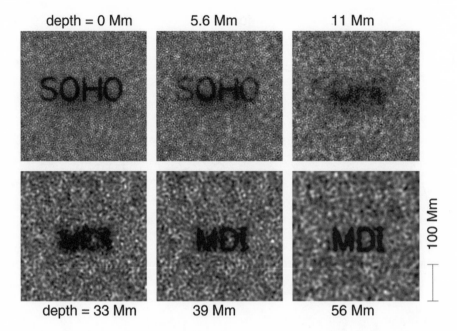

FIG. 8.11 In a numerical simulation of acoustic holography, two subsurface absorbing objects are viewed in sound waves. As the viewing depth is changed, the object comes into sharp focus.

changed the depth to which they focused (see the panels in each column in fig. 8.11), the absorber became sharper or fuzzier, just as with an ordinary camera.

Next they had to work out how to use the waves they could observe at the front side of the Sun to detect spots and active regions on the far side. Figure 8.12 illustrates how this is done. An absorbing sunspot, located on the far side, scatters the traveling waves that make up the five-minute oscillations. A bundle of scattered waves travels deep into the Sun, spreading apart as it goes, bounces off opposite sides of the Sun, and continues on toward the front side. There the waves arrive in a circular band that faces the observer, who has to record the oscillations within the band for a fairly long time. Then, after the observation, the observer computes the ray paths the waves must have taken to arrive at the front side. When the waves are projected backward, a picture composed of sound waves of the scattering object is obtained, in this case a dark sunspot.

Now, with this method, acoustic holography, it is possible to image not merely

focus

solar surface

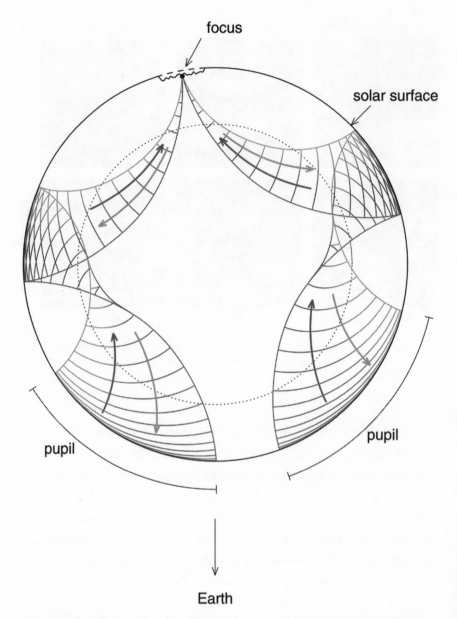

pupil

pupil

Earth

FIG. 8.12 The ray paths of sound waves that are used to view the far side of the Sun. Long wavelength waves plunge deep into the sun, loop back to the surface once and land in a circular band facing the observer. Observations of these waves enables the observer to reconstruct the far-side structures.

a single sunspot but a large active region on the back side of the Sun. The first pictures like these were constructed from data from the MDI instrument aboard SOHO in 2001, and were announced with considerable excitement by NASA and the European Space Agency. Figure 8.13 compares the sonic images of a region on the back side with visible images of the same region after it had rotated to the front side of the Sun. The arrows point to a large sunspot, and another chain of spots is visible on the right side of the images.

The holographic image is coarse by comparison with the usual optical image

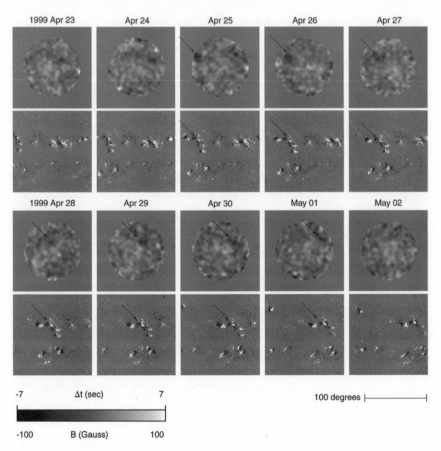

FIG. 8.13 A comparison of sonic images of the back side of the Sun (top row) and visible images of the same regions after they rotated to the front side (bottom row). The period covers ten continuous days. The arrows point out some corresponding magnetic features in the images.

for several reasons. First, only a few special sound waves are able to make the double bounce we see in figure 8.12. Second, the sound waves may be scattered by other obstacles on their trip to the near side. And finally, the region on the back side takes at least ten days to rotate to the front side and may change shape and size during that time.

Nevertheless, this is a wonderful achievement, with several practical applications. Solar active regions are the breeding grounds of flares. These violent explosions spray the Earth with powerful X rays and electrically charged particles, emissions that can cause serious damage to satellites or space-walking astronauts. Therefore, the governments of the United States and its allies are vitally interested in getting advance warning of the appearance of a large active region.

Similarly, large chunks of the solar corona tear off the Sun at least once a day. If a billion tons of this ionized gas happens to reach the Earth, it can interact with the Earth's magnetic field. As a result, large electrical currents can be induced in transmission lines and oil pipelines, causing extensive damage to sensitive electrical equipment. Power and oil companies are therefore also interested in forecasting solar activity.

Holographic seismology has become a minor industry for Doug Braun. Images of the far side of the Sun are computed almost daily now from SOHO data and are posted on the World Wide Web. The technique has become as familiar to solar astronomers as a weather forecast.

ROTATION, CONVECTION, AND HOW THE TWAIN SHALL MEET

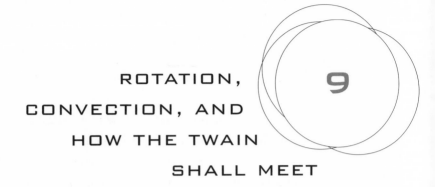

9

YOUNG THEORISTS, just starting their careers as postdoctoral scientists, are often assigned tiny cubicles at their new places of employment. One such fledgling I knew covered the gray wall in front of him with a huge poster of the Sun's internal rotation (see fig. 9.1). This road map to the interior, dated 1997, was extracted from two months of observations by the MDI instrument aboard the SOHO satellite. He hung the map there to keep reminding himself of two important goals. Should he accomplish either, his reputation would be made. The first goal was difficult but relatively straightforward to reach. It was to extend the rotation map into the innermost core of the Sun, inside a fractional radius of 0.3, where the energy of sunlight is liberated. Recall that the *temperature* in the core (and therefore the neutrino production) has been pinned down very recently, but the *rotation* in the core is still an open question. What lurks in these unexplored depths? Can the Sun be hiding a large fraction of the spin it was born with?

The second task was by no means straightforward. It was to demonstrate, from basic physical principles, just how the observed pattern of differential rotation in the convection zone arises. Such a task would involve a series of horrendous calculations with supercomputers. Even worse, it would require entirely new insights into the physics of convection, because convection and rotation are intimately related in the Sun.

That youngster is fortunately not alone in his ambitions. Half the world population of helioseismologists has been focusing on these two tasks for the better part of a decade. In this chapter we'll see how far they've come.

PUSHING TOWARD THE CENTER

Take a look at figure 9.1, a map constructed in 1997 from MDI data. The convection zone of the Sun is spinning in a complicated pattern that varies in both depth and latitude. As we proceed inward at the equator (latitude zero degrees), the angular speed increases slightly just below the surface, remains constant for some distance, and then takes a steep dive just below the base of the convection zone (0.7R).

At the high latitude of 60 degrees, the angular speed at the surface is lower than at the equator, actually dips a bit, and then rises steeply to join the equator's speed at the bottom of the convection zone. At an intermediate latitude, such as 30 degrees, the angular speed behaves more like the equator's but with a definite slow decrease throughout the convection zone. Finally, the radiative zone, between 0.6 and 0.4 radii, rotates as a solid body at an angular speed about the same as that at latitude 30 degrees.

The map ends at a radius of 0.4R. Nothing certain was known in 1997 about the rotation of the core. Moreover the region at or around the base of the convection zone, where the rigidly rotating radiative zone meets the differentially rotating convection zone, was a bit of a jumble. As we will see, this transition zone, the "tachocline," has been studied intensively in the past five years because it is probably where the solar cycle originates.

The biggest immediate challenge, however, was to push deeper into the core of the Sun, where solar energy is generated. Astronomers wanted to determine how fast the core rotates, how much of its original angular momentum the Sun still retains there. That could reveal how much turbulent mixing occurs in the core, a factor that affects the evolution of the Sun.

Astronomers have learned that the surface rotation of stars decreases sharply once hydrogen begins to burn in their cores. However, some models of stellar evolution suggest that the Sun's core could still be rotating ten to fifty times as fast as the equator at the surface. In fact, the earliest observations by the Bir-

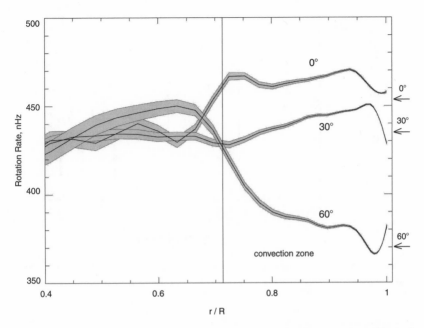

FIG. 9.1 Variation of the Sun's angular rotation rates at different depths and latitudes. The shaded bands indicate the probable errors. For reference, a rate of 430 nanohertz corresponds to a rotation period of 26.9 days.

mingham group seemed to confirm such high speeds. As more data accumulated, the estimated speeds declined. As of 1994, the actual speed was still uncertain.

Improving estimates of the core's rotation posed a daunting observational problem. The only p-modes that penetrate the core are of lowest degree (such as $L = 1, 2, 3$), and they have very few associated azimuthal modes. An oscillation with $L = 2$, for example, has only five azimuthal modes ($M = -2, 1, 0, 1, 2$). This dearth of allowed modes severely limits the information one can glean about the core's rotation. What's more, no instrument was able to resolve the spatial patterns of the azimuthal modes over the solar disk until the LOWL instrument went into action. The only other way to probe the core was to determine the frequencies of the low-L modes more accurately, and that would take years of continuous observations.

The Birmingham group, BISON, boasts one of the longest-running series of observations, starting with one station and ending with six. In 1996 William

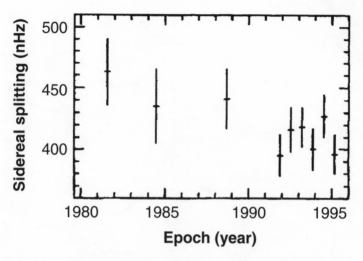

FIG. 9.2   The rotation rate of the Sun's core, derived from BISON data (1981–1995). The average rate over this fourteen-year interval corresponds to a rotation period of 27.9 days.

Chaplin and friends analyzed batches of their data from 1981 to 1995. Remember that the BISON instrument (a resonant scattering cell) records a surface average of the five-minute oscillations, with a heavy weighting of those "sectoral" modes that peak at the equator. This means they could derive only a crude average rotation rate for the core as a whole.

Figure 9.2 shows how their estimates of this rate varied over the fourteen-year period. The average frequency of rotation was 415 nanohertz, close to the equatorial rate at the surface. Their results certainly ruled out a core rotating two or three times as fast as the surface. But other observers had derived mean core rates as high as 452 nanohertz. Who was right?

By 1996 Steve Tomczyk's LOWL instrument in Hawaii had accumulated a year of data. Recall that this device was especially designed to record all L- and M-modes from zero to 80 and to resolve the azimuthal modes on the solar disk. Its stability and precision make it one of the preferred instruments for probing the deep interior even though it is not part of a network.

In figure 9.3 we see the angular rotation rates, extending down to a radius of

0.2R, which Tomczyk and coworkers derived from their data. In the radiative zone (radius 0.6R to 0.3R), the curves from all latitudes overlap, confirming a solid-body rotation of about 430 nanohertz. The results become very uncertain below a radius 0.3R, however, so the steep decline of angular speed there may not be real.

Thierry Corbard, a bright young Frenchman, repeated Tomczyk's analysis after an additional year of LOWL data had accumulated. He and his colleagues also found a mean rotation rate of 430 nanohertz in the radiative zone, now with a slight latitude variation around a radius of 0.4R.

Could a better scheme of analyzing the data squeeze out more information? In 1998, Paul Charbonneau from the High Altitude Observatory, Jesper Schou from Stanford University, and Michael Thomson from Queen Mary College in London decided to try out a sophisticated technique called "genetic forward modeling." This procedure for finding an optimum solution to a fuzzy problem was inspired by the biological process of evolution by natural selection (note 9.1).

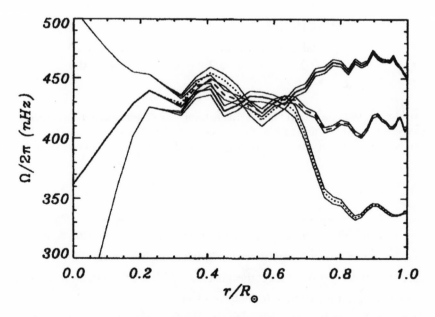

FIG. 9.3   One year of LOWL data was analyzed by Stephen Tomczyk and colleagues in 1996. They were able to extend the rotation profiles down to a radius of 0.3R, but not deeper.

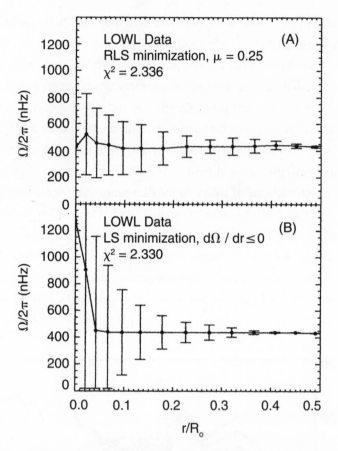

FIG. 9.4 With two years of LOWL data in hand, Paul Charbonneau and associates used genetic modeling to determine the rotation rate of the core, down to about 0.1R. Panels (A) and (B) show consistent results from two different methods of inversion.

Basically, one allows the set of parameters involved in the problem to mix and match ("breed") over many successive "generations." A suitable "fitness" test is used in each generation to select the most favorable variations in the parameters, so that an optimum solution is finally reached.

It sounds complicated and it is. Charbonneau and company made many preliminary tests on artificial data with realistic random errors to learn how to drive this mathematical juggernaut. Figure 9.4 shows their best efforts on two years of

LOWL data. The rotation rate is flat all the way down to radius 0.1R and is consistent with *rigid* rotation (at 424 nanohertz) below radius 0.5R.

All this is fine except for the issue of *systematic* errors. (Your chronometer may lose only a second a month but show the wrong day.) The LOWL is only one instrument, not part of a network, and is therefore subject to a powerful resonance introduced by the day-night cycle. Tomczyk had worried about this two years earlier and showed this effect is not important in his data sets. But there is also the problem of "leakage," in which a real oscillation frequency coincides accidentally with a spurious frequency left over from the day-night cycle. Only a comparison with other independent data sets could reveal the size of this kind of systematic effect. Fortunately such a data set was ready by 1998.

The GOLF instrument aboard the SOHO satellite is basically a resonance scattering cell (see note 4.2) that uses two spectral lines of sodium in sunlight to measure Doppler velocities. The device has no spatial resolution over the solar disk but is very stable, and, perched aboard SOHO, it sees the Sun twenty-four hours a day. Like the BISON instrument it emphasizes sectoral modes that peak at the solar equator.

By 1998, GOLF had observed the Sun continuously for a year. Thierry Corbard and a team of coworkers took on the job of combining the GOLF low-L data with MDI medium-L observations to derive the rotation rate in the core. Figure 9.5 displays one of their results. The rotation rate is flat at about 430 nanohertz from 0.6R to 0.3R and then explodes to 550 nanohertz below 0.2R. When the same analysis was carried out on GONG data, the flat portions agreed, but the rotation rate inside 0.3R *plunged* to values as small as 370 nanohertz. Clearly, neither data set was adequate to probe below 0.2R.

Bill Chaplin and his friends may have had the last word, at least until March 1999. They combined thirty-two months of BISON data with two years of LOWL observations and tried out three different inversion schemes. None of these schemes was able to resolve latitude variations of rotation below 0.25R, simply because the data do not contain sufficient modes for the purpose.

So Chaplin and associates limited their goal to answering a simple question: Does the rotation increase or decrease inside a radius of 0.25R? One inversion indicated a downturn (fig. 9.6A), but with a 50% chance there might be an up-

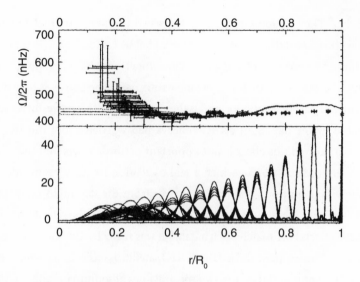

FIG. 9.5 Thierry Corbard and coworkers determined the average rotation of the core from one year of continuous GOLF observations. The rates are reliable down to a radius of 0.3R. The lower panel shows the "kernels" used to isolate different depths. Note that they become weaker and broader in the core.

turn. However, this team places greater trust on the results of a second inversion, shown in figure 9.6B. The filled circles also indicate a downturn toward slower speeds in the core, but the error bars tell the same old story: a big uncertainty. The most likely result, they concluded, is that the core (from 0.15R to 0.3R) rotates as a solid body at the same rate (435 nanohertz) as the radiative zone.

Further progress in pressing toward the center of the Sun may wait for somebody to detect gravity waves. Theorists predict that these waves (driven by buoyancy forces, not pressure gradients) penetrate the core, but they also predict that the waves don't reach the surface. Observing them will be difficult, if not impossible.

But stay tuned. Theorists are not infallible.

THE HOLY GRAIL

Perhaps the most challenging problem still facing solar physicists is to explain in physical terms how the sunspot cycle arises. Sunspots come and go in an eleven-

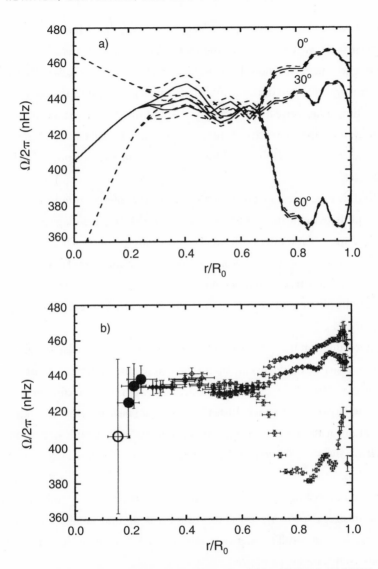

FIG. 9.6   In 1999, William Chaplin and colleagues combined BISON and LOWL observations to derive the rotation of the core. Panel (a) shows reliable results to 0.2R from one inversion scheme. Panel (b), from another scheme, shows a possible downturn in the rates, but the results are also consistent with a flat curve (rigid rotation).

year activity cycle, and many other features of the Sun, such as the corona and the solar wind, vary in step with the spots. For many years, theorists have been working to understand how this complicated magnetic cycle arises. They recognized very quickly that the basic machinery must lie in or near the bottom of the convection zone, as we'll see in chapter 10, and that the ultimate driving forces of the cycle are the convective motions. For a complete explanation of the physics of the solar cycle (the Holy Grail) they needed to understand convection in a rotating star.

We now have a reasonably accurate map of the internal rotation of the Sun, except perhaps in the innermost core. Long before helioseismology provided this splendid map of the interior, however, some enterprising solar physicists were trying to understand how the *surface* pattern of rotation arises. They thought that if they could construct a numerical model of the convection zone that correctly predicts the differential rotation at the surface, they could use it to investigate the sunspot cycle.

Peter Gilman at the High Altitude Observatory and Gary Glatzmaier (now at the University of California at Santa Cruz) were among the leaders in the field during the 1980s. They agreed that the problem was much too complicated to be solved with paper and pencil. Instead, they would have to resort to a numerical simulation of the process, using fast computers. So they built huge computer codes to solve the equations of hydrodynamics, in three dimensions, in the rotating Sun.

In their description of convection, giant convection cells (a tenth of the solar radius in size) rise slowly, spread out, and break up into smaller cells, which in turn break up into still smaller cells, and so on. At the top of the zone, the cells radiate away their heat and sink.

If the Sun didn't rotate, the forces of gravity and buoyancy, which drive the cells, would act the same anywhere in the Sun because these two forces are simply radial. In that case, the forces would be perfectly symmetrical, and there's no way the cells could transport angular momentum from one latitude to another. Hence, there would be no differential rotation in the Sun.

Because the sun *does* rotate (with an average period of about twenty-seven days), the Coriolis force comes into play. It tends to tilt the largest cells away from

FIG. 9.7   In early models of global convection, the flows were treated as nearly laminar. In such cases, giant "banana" cells appear as neighboring north-south rolls. Solid contours indicate upward velocities, and dotted contours indicate downward velocities.

the radial direction and also away from the north-south direction. These tilts combine in such a way as to transfer momentum across parallels of latitude. In Gilman and Glatzmaier's simulations, the giant cells lie in north-south rolls (fig. 9.7) and because of their appearance are called "banana cells." With this kind of model, Gilman and Glatzmaier achieved a remarkable success: they were able to reproduce the observed variation of rotation speed with latitude on the Sun's surface (see fig. 5.3).

Their codes could then predict the pattern of rotation inside the Sun. They found that the angular speed of rotation was constant on cylindrical surfaces that were centered on the Sun's rotation axis (see fig. 5.2). This result implied that the angular speed *decreases* the deeper you look inside the sun, which has a most undesirable consequence. If this pattern of rotation really exists, sunspots would appear at higher and higher latitudes on the rise of a solar cycle, which is exactly opposite to what one observes.

Two experts had independently arrived at the same incorrect answer. What had gone wrong? They had used the most detailed description available of all the physical processes and still missed the mark. In their postmortems they realized that they hadn't adequately described the *turbulence* of convective flows. And that was for a very good reason. Turbulent flows, as we shall see, break down into a huge range of sizes. Some parts of the flows are coherent over the whole depth of the convection zone, while others approach molecular sizes. No supercomputer of the 1980s could begin to cope with that huge range. What was not obvious, in retrospect, was that turbulent flows are *qualitatively* different from smooth flows.

Solar physicists would have to learn much more about turbulent convection before they could attack the problem of differential rotation and its role in the sunspot cycle. To understand what they have learned, we need to look back a bit.

HOW IT ALL BEGAN

In 1901, Henri Bénard, a French physicist, was investigating how different liquids would behave when he heated them in a shallow pan. He experimented with water, oils of different viscosity, and paraffin. At first, he could see very little flow in the fluid; heat was simply being *conducted* from bottom to top, as though the liquid were really a solid. Then, as he raised the temperature, he noticed a distinct change in behavior. A heavy oil would form a stable surface pattern of polygonal cells, in which hot oil would rise at the center, and cool oil would sink at the edges. The oil was now *convecting* heat from the bottom of the pan to the top and dumping it into the cooler air near the surface.

Lord Rayleigh, the famous British physicist, became interested in Bénard's cells. In 1916 he set up a simple mathematical model to interpret the experimental results. He learned that the onset of convection in a fluid like oil depends on such things as its depth, viscosity, and heat conductivity, as well as the temperature gradient from top to bottom. He combined the essential parameters of the situation into a dimensionless number, now known as the Rayleigh number, which determines under what conditions a liquid or gas will begin to con-

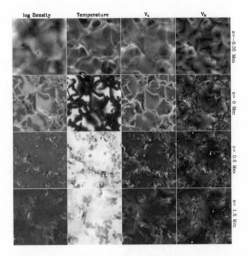

FIG. 9.8   The properties of convection at different depths near the solar surface. The columns show gas density, temperature (white is hot, black is cold), vertical velocity (white is downward), and horizontal velocity (black is diverging, white is converging). At z = 0, the actual surface, the temperature map resembles granulation. Downflows appear in the lanes and in long fingers that reach the bottom of the layer (z = −1.5 Mm).

vect heat. For corn syrup, the critical Rayleigh number is about 10^5, for the Sun about 10^3.

Bénard's pretty cells were examples of smooth "laminar" flow in a convecting liquid. But when Bénard turned up the heat sufficiently, the cells broke down and the liquid began to roil in chaotic, *turbulent* flow, changing its pattern from moment to moment. Osborne Reynolds, another famous British physicist, devised his own dimensionless number, which determines the necessary conditions for a gas or liquid to switch from laminar to turbulent flow. In common fluids like air and water, the critical Reynolds number is about 1,500.

The surface of the Sun is covered with a pattern of fairly stable convection cells ("granules"; see fig. 1.1) that look a lot like Bénard's cells. They last for about ten minutes and have internal speeds of about a kilometer per second, much less than the sound speed. So you might think that all the flows in the convection zone are laminar. Indeed, all the early work on convection in the Sun was based on the idea of discrete cells in which the flow was laminar.

For the deeper layers, however, that was a very poor description. The Reynolds number in the middle of the convection zone is about 10^{12}, a *billion* times larger than the critical value. That means the flows in the middle are incredibly turbulent and the convection must be very different from near the surface. In addition, the Rayleigh number in the convection zone is about 10^{22}, far beyond the critical limit, which means the convection is also extremely vigorous.

An enormous amount of laboratory research has been done on convection since Bénard did his famous experiments, and physicists have learned a lot about turbulent convection. Solar physicists have had a terrible problem in trying to apply these laboratory results to the Sun, however, because the Sun's convection is so extreme. As bigger and faster computers have become available, however, several groups around the world have been able to carry out numerical experiments that simulate, to a very limited extent, how convection works in the Sun. We will focus on a couple of these, with apologies to the rest.

THE SUN IN A BOX

Åke Nordlund, now at the Copenhagen University Observatory, was one of the first to try to simulate solar convection. He focused on the top few thousand kilometers of the convection zone where the solar granules appear, because this layer seemed the easiest to explain and because, by the mid-1980s, observations of granules had improved tremendously. Several observatories, including the German station at Izana and the Sacramento Peak Observatory in New Mexico, had obtained startlingly sharp images of granules and had followed their evolution and motions as well. The time was ripe.

Nordlund developed a computer code to describe the behavior of gas in a rectangular box, a few thousand kilometers wide and deep. The box was intended to represent a typical piece of the upper convection zone. He divided the box into a three-dimensional grid of points. Solving his hydrodynamic equations, he could calculate all the properties of the gas, including its velocity, at each of these points, for each moment of time.

A steady flow of heat was allowed to enter the bottom of the box and allowed to radiate into space at the top. To begin with, the gas was static and stable, simply resting, but as the heat entered the box, the gas began to churn. After a while, discrete cells formed and filled the box. They would appear and disappear randomly, in simulated times of a few minutes, just like real granules.

The acid test was to compare the shapes and size distribution of the simulated granules with the best observations. And indeed, the comparison was quite good. A typical artificial granule was about 1500 km across, with a bright hot center and dark borders and a lifetime of about ten minutes. But the really interesting result lay beneath the granules. According to the contemporary ideas of solar convection, heat was supposedly carried to the Sun's surface by a kind of bucket brigade of cells. Large cells at the bottom would pass their heat to smaller cells higher up. When the cells at the top cooled, they were supposed to sink gradually, in broad diffuse flows.

Instead, Nordlund's simulations showed clearly that the gas descended *rapidly* in narrow, twisting "fingers," something like water draining from a bathtub. These fingers (or "vortices") originated in the cool dark lanes between granules and plunged down to the bottom of his numerical box. Physicists had seen this sort of behavior in the laboratory, when a cool dense liquid on top of a warm light liquid becomes unstable and collapses in narrow twisting fingers (see note 9.2 on the Rayleigh-Taylor instability). Here was food for thought: the old picture of convection needed fixing.

In the late 1980s, Nordlund teamed up with Bob Stein, a young solar theorist at Michigan State University. Stein had been working independently on solar convection and had some fresh ideas. With new simulations that extended as deep as 2500 km, they investigated how the hot rising gas and cool descending vortices were interconnected. By 1989 they were able to report some novel results.

Figure 9.8 shows how the temperature, density, and vertical and horizontal velocity vary throughout the depth of their box. At the surface (labeled $z = 0$) the temperature and velocity maps resemble the familiar granules. But these are shallow and fade out only 500 km under the surface (at $z = 0.5$ Mm). In their place, we see the narrow twisting downflows (colored black in the velocity map and white in the helicity map), extending down from the intergranular lanes, and through the slowly rising hot gas. So the flows are *asymmetric*, unlike any that had been expected previously.

These simulations confirmed that the flows in the granules are laminar, but the descending vortices were turbulent without a doubt. Unfortunately, in 1989 Stein and Nordlund were not able to describe the turbulence. The problem, still faced by all investigators of convection, is that turbulent flows have a tremendous range of sizes, all the way down to molecular sizes, and change rapidly. To model all this fine structure and rapid change, even approximately, would require far more computer power than anyone had until the late 1990s. At that point, computers could not extend these detailed studies below a depth of about 2500 km, a mere one percent of the depth of the convection zone. And, more important, nobody knew if these pretty simulations had any relevance to the fierce convection in the real Sun. Then, in 1996, a glimpse of the real world arrived from the helioseismologists.

A SLICE OF THE REAL SUN

Tom Duvall, you will remember, had invented a new method of probing the interior of the Sun with sound waves. He called it the "time-distance method" or "solar tomography," and we described his first uses of this technique in chapter 8. By 1996, the Michelson-Doppler instrument aboard the SOHO satellite had accumulated a long, continuous record of the five-minute oscillations over a large part of the solar disk. So the time was ripe for Duvall to explore convection. He teamed up with Alexander Kosovichev, a senior research scientist at Stanford University. Kosovichev is a talented theorist who has worked on all sorts of solar problems, including convection, magnetic fields, and the origin of the five-minute oscillations. He had also suggested an improvement to Duvall's original time-distance method. So these two took on the tedious job of analyzing enough data to reveal convective motions beneath the solar surface.

Figure 8.7 shows their results graphically for a vertical slice of the convection zone 150,000 km wide and 5000 km deep. The arrows show the direction of the flows, and the colors indicate regions of hotter (red) and cooler (blue) temperatures (see color gallery). These first results show some similarity to the Stein-Nordlund simulations but also some differences. Most of the observed flows are weak and horizontal, in agreement with the simulations. But in the coolest parts, near the surface, the observed flow is *diverging,* not converging or sinking, as the simulations suggest. Moreover, although narrow vertical plumes are present, some flow down (e.g., on the left of fig. 8.7), others flow up (in the center), and some up and down flows are side by side. Is the convection in the Sun more com-plicated than simulations indicate, or do the existing observations lack sufficient spatial resolution? Perhaps both. Only time will tell.

PUTTING A SPIN ON IT

Juri Toomre, a professor of astrophysics at the University of Colorado, leads a team of intrepid explorers into the hazardous jungles of stellar convection. For over two decades he and his colleagues have been struggling to understand con-vection in a rotating star like the Sun. Lately, with the help of powerful comput-ers, they've made substantial progress.

Convection, especially turbulent convection, is difficult to understand even without the complications that rotation introduces. But the Sun rotates in a pe-riod comparable to the lifetime of the largest flows in the convection zone, which means that rotational forces will change the flows radically. Nobody has a com-plete understanding of this messy coupling between rotation and convection, which ultimately produces the differential rotation observed by the helioseis-mologists. It is a riddle, wrapped in a mystery, inside an enigma, as Winston Churchill might have said.

To tackle the problem Toomre and his postgraduate helpers have broken it down into three parts. First, they investigated how turbulent convection in an imaginary box depends on such properties as the Rayleigh and Reynolds num-bers. Fausto Cattaneo, Neal Hurlburt, and Nicolas Brummell were key players in this study. They were able to confirm much of what Stein and Nordlund had found, and with much more detail. Next, they studied how centrifugal and Cori-olis forces, produced by rotation, modify the convective flows in an isolated box. And finally, they put all these ingredients into a spherical rotating shell to see whether they could predict the observed pattern of differential rotation.

To do all this, they needed to include as many fine details of the turbulent flow as possible until they ran out of computer power. Only then could they ask mean-

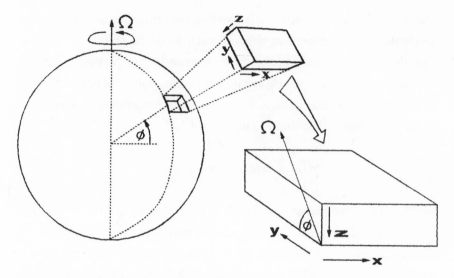

FIG. 9.9   N. Brummell and coworkers investigated the effects of rotational forces on convection in a box. The boxes were selected at different solar latitudes. In the box on the right, the vector Ω points in the local direction of solar rotation.

ingful questions. What is the topology of the flow? Which sizes of turbulent flows carry the most heat? Which sizes carry the most momentum or angular momentum? How does the convective flow depend on the speed of rotation or the stickiness of the gas?

Toomre and company use some heavy artillery in this battle. Their high-quality science has won them access to some of the most advanced supercomputers in the world such as the Cray T3E, located at national centers in San Diego, Pittsburgh, and Los Alamos. These huge machines allow them to follow the evolution of flows at a million discrete points in a box, compared to Nordlund's original 50,000 points.

Even this state-of-the-art technology is insufficient, however, to follow the smallest details of convection over the full depth (250,000 km) of the convection zone and over the full range of flow lifetimes. So Toomre and coworkers have learned to compromise. They could omit or fudge some features of convection, but others are essential and had to be treated with some degree of realism.

First, they had to allow the gas to be compressible and move in three dimensions. Then, the hundredfold increase of density, from the top to the bottom of the convection zone, had to be taken into account. Also, the Sun's relatively rapid rotation required them to allow for twisting flows. And finally, to examine truly turbulent flows they had to reduce the viscosity of the gas.

They began with a simple slab, extracted at some latitude from the top of the Sun's convection zone (fig. 9.9). Depending on its latitude, the slab experiences stronger or weaker centrifugal and Coriolis forces. At the Sun's poles these forces vanish, while at the equator they reach maximum strength.

As the team decreased the viscosity of the gas, the flows became more turbulent, and vortical structures of all sizes and directions filled the whole volume of the slab. In the midst of all this chaos, a few long-lived vortices extended vertically through the full depth of the box. Moreover, the now-familiar cellular pattern, resembling the solar granulation, always survived at the top of the box.

One of the biggest surprises to come out of these studies was the way rotation affects the long-lived vortices. These tight whirlpools tipped away from the vertical direction and *aligned themselves* with the local direction of rotation (fig. 9.10). The team was able to explain this curious behavior in terms of the Coriolis forces acting on the vortices. They immediately recognized that they were seeing a new mechanism for the transport of momentum and angular momentum from one latitude to another. In such a tilted turbulent vortex, the horizontal and vertical motions are *correlated;* that is, as the gas spirals downward, the diameter of the vortex shrinks. Such correlations could feed back through the nonlinear equations to modify the Sun's average rotation and in this way produce the differential rotation. The only way to find out was to build a toy convection zone.

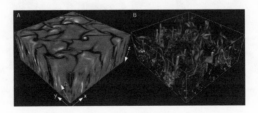

FIG. 9.10   In panel A we see a 3-D snapshot of the convection in a box (black areas are cool and downflowing). Turbulent downflowing vortices in panel B tend to align with the local direction of rotation, Ω.

PUTTING IT ALL TOGETHER

Julian Elliott, a former student of Douglas Gough at Cambridge University, joined Toomre's group in 1997, rolled up his sleeves, and tackled this brute of a simulation. He used the hottest computers and got lots of advice from the senior scientists, Toomre and Glatzmaier.

The team's basic numerical model is a three-dimensional shell that is filled with a perfect gas. As usual, the gas is heated from below and loses heat out the top. But to avoid having to model the tiny granules at the same time as the larger flows, they cut off the model at a fractional radius of 0.96, just below the granulation. Also, to avoid having to follow transient sound waves throughout the simulation, they used a form of the hydrodynamic equations introduced by Douglas Gough (the "analastic approximation") that preserves slow buoyancy forces while suppressing fast pressure changes.

The computed patterns of convection are very complicated and change over a few days. Figure 9.11 shows two snapshots of the radial velocity near the sur-

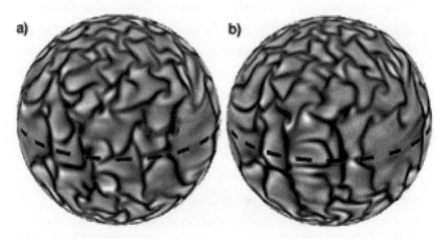

FIG. 9.11 Two snapshots, ten days apart, of the radial velocity near the top of the convection zone, from a recent, low-turbulence simulation. Bright tones indicate rising motions, dark tones indicate downflows.

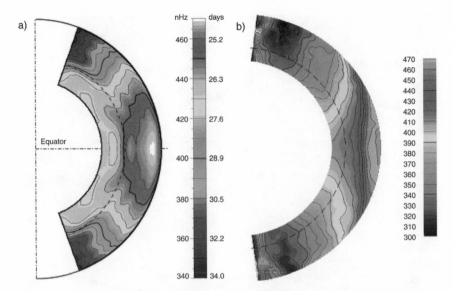

FIG. 9.12 A comparison of observed (a) and simulated (b) patterns of differential rotation. The angular speeds are indicated with a color code, red corresponding to rapid rotation and blue to slower rotation (see color gallery). The match is quite good, except for the remaining cylindrical pattern (the parallel vertical contours) in the simulation.

face, separated by about ten days. In this simulation the level of turbulence was purposely reduced, and the resulting pattern looks something like the surface of the human brain.

In order to compare their simulations with the helioseismic results, the team averaged the computed flows over three rotation periods (about eighty-five days). Their best effort, as of 1998, is compared with the helioseismic map of rotation in figure 9.12.

For the first time, they were able to reproduce two important features of the Sun's differential rotation, namely that the rotation is nearly constant on *radial* lines at midlatitudes, and that the rotation rate at the surface decreases from the equator to the poles.

The team learned quickly that what you get depends on what you put in. Not surprisingly, the results are sensitive to the parameters you choose (like the Reynolds number) and the boundary conditions you impose. Marc Miesch, an-

FIG. 9.13 A snapshot of turbulent convection in a box, computed at a *billion* points. This 1999 simulation at the Lawrence Livermore National Laboratory had ten times the spatial resolution attained at that time by solar physicists. The image shows tangling vortices and vortex sheets at one instant.

other young postdoc, explored these factors in collaboration with the Toomre team and showed that one can improve the match with observations with a slight change in the treatment of the top of the convection.

What can we conclude from all this effort? First, despite all the progress in modeling turbulence in a box, the simulations of the full convection zone are still limited by computer power. As of 2000, the simulations of Toomre's group were still too viscous to completely eliminate a tendency toward cylindrical rotation. Second, the more computer power you have, the more complex the flows become. See, for example, a snapshot from a turbulence study at the Lawrence Livermore National Laboratory (fig. 9.13), made with ten times the spatial resolution Brummell used (fig. 9.12).

Finally, the top of the convection zone is even more critical than previously supposed, because the all-important downflowing vortices arise there. That makes studies like those of Stein and Nordlund more urgent. Recently they have improved their simulations and are seeing tangled vortices galore.

It appears that we will have to wait for more patient work to be completed before we can say we fully understand why the Sun rotates the way it does. But that hasn't discouraged attempts to understand the origins of the solar cycle, as we shall see in the next chapter.

10   THE SOLAR DYNAMO

TAKE A LOOK at figure 10.1, which shows a typical view of the million-degree corona over the disk of the Sun. These bright loops and threads are produced by strong magnetic fields that poke up through the surface. Hot glowing gas, trapped in the loops, outlines the shape of the field, which would otherwise be invisible.

All the forms we see in the corona, including the gigantic streamers, owe their existence to the presence of magnetic fields that the Sun produces, somehow, deep inside its body. Solar physicists are working hard to explain just how a low-tech ball of gas like the Sun can generate magnetic fields. New observations of the solar interior, provided by the exquisite tools of helioseismology, are guiding that quest. In this chapter we will learn how the campaign is going. But first let's recall what past observations of magnetic fields have revealed.

SPOTS IN YOUR EYE

George Ellery Hale, an American astronomer, was the first to discover that the Sun is a magnetic star. From 1904 to 1923, Hale was the director of the Mount Wilson Observatory and probably the most prominent solar astronomer in the world. He was a man of tremendous energy, talent, and enterprise and founded

FIG. 10.1 These beautiful loops are formed by the magnetic field in the Sun's corona. The loops are filled with plasma at a temperature of about a million kelvin.

and directed three great observatories, each the foremost of its time: the Yerkes Observatory, the Mount Wilson Observatory, and later, the Palomar Observatory with its 200-inch telescope. He was the driving force in the creation of the *Astrophysical Journal*, the American Astronomical Society, and the International Astronomical Union. As strategist, fund-raiser, and propagandist, Hale was without peer.

Hale also found time somehow to pursue his own research. In 1890 he invented the spectroheliograph, independently of the French astronomer, Henri Deslandres. This is a device for viewing the chromosphere, the layer that lies just above the surface. When viewed in the light of a particular spectral line of hydrogen (H alpha, at 656.3 nm), the chromosphere around sunspots would often display a characteristic spiral pattern of dark threads (fig. 10.2).

Hale was intrigued. He guessed that the spiral indicated that the spots were rotating. If they were, classical electrodynamic theory suggested they might produce magnetic fields, so he decided to look for such fields. He used a property of atomic spectra that Pieter Zeeman, a Dutch physicist, had recently discovered— namely, that spectral lines would split when atoms were immersed in a magnetic

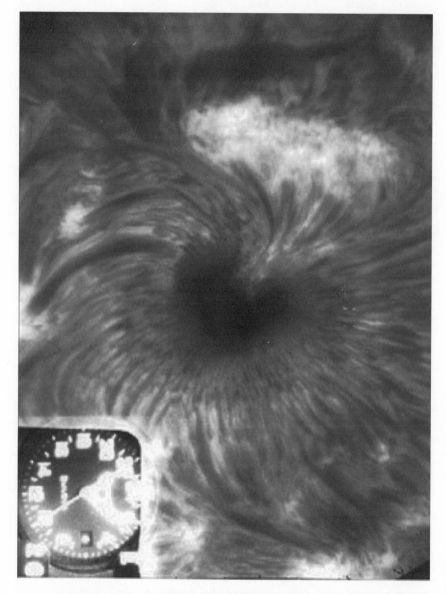

FIG. 10.2 Some sunspots have a spiral pattern of magnetic fibrils, with some tendency to be counterclockwise in the Northern Hemisphere and clockwise in the Southern Hemisphere.

field. Using a powerful spectrograph at the eighteen-meter tower telescope at Mount Wilson, Hale searched for split spectral lines in sunspots.

And lo! He found them! From the size of the splitting he could estimate that a large sunspot has a field strength of 2000 to 3000 gauss, or several thousand times as large as the Earth's field at its magnetic poles. The sunspots appeared in pairs of opposite magnetic polarities, with a "negative" field directed inward in one and a "positive" field directed outward in the other. Hale was wrong about the origin of the sunspot fields, but that doesn't detract from his groundbreaking discovery. There's nothing disreputable about serendipity in science. Hale went on to discover how the magnetic properties of sunspots change during the well-known sunspot cycle. An obscure pharmacist in Germany had discovered this cycle back in the 1800s.

The Pulse of the Sun

Samuel Heinrich Schwabe was a character very different from Hale. The eldest of ten sons, he lived all his life in the town of Dessau in eastern Germany. Schwabe was a successful pharmacist by profession but an amateur astronomer by choice. He began observing the Sun systematically in 1826, with a telescope he had won in a lottery, and continued with a fine instrument he ordered from the Fraunhofer workshop in Munich. Every day the sky was clear, he recorded the number of sunspots he could see.

After two years he noticed a significant increase in the numbers. A cautious man, he decided to continue observing before making any announcement. The numbers continued to increase and then, around 1831, to decrease. Schwabe was a paragon of patience. He kept at his task for *seventeen* long years, and finally, in 1843, announced his discovery of a ten-year cycle.

At first he was disbelieved. Only after several professional astronomers, including Rudolf Wolf, confirmed his discovery were his efforts appreciated. Wolf established an average length of the cycle as 11.1 years, but the length can vary from 9.5 to 12.5 years. Longer cycles are now known, such as the Gleissberg cycle of about eighty years.

Richard Christopher Carrington was the next to make a fundamental discovery about the behavior of sunspots. He was the son of a successful English brewer

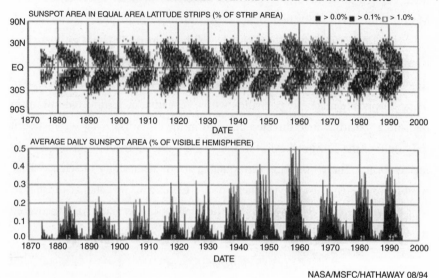

FIG. 10.3 Maunder's famous "butterfly diagram" (top panel) shows how new sunspots appear at lower and lower latitudes during a solar cycle. The lower panel shows how the number of spots varies in a cycle.

and inherited his father's fortune. With plenty of money and all the leisure time in the world, Carrington devoted himself to solar astronomy. (Now there's a serious person for you!) He began observing sunspots systematically in 1853 and by 1861 had a definite result. He determined that the spots of a new cycle appear first at high solar latitudes, and as the cycle progresses they appear at lower and lower latitudes, until finally after eleven years the last few are born within a few degrees of the solar equator. Incidentally, Carrington's careful observations also revealed that the equator of the Sun rotates faster than higher latitudes, the now-famous "differential rotation" of the Sun's surface.

Ernest Maunder, a British astronomer, was the first to display Carrington's results in a now-famous "butterfly diagram." In figure 10.3, a modern diagram, you can see how the latitudes of new spots changed periodically in cycles dating back to 1880. Maunder also rediscovered an even more interesting result: from approximately 1645 to 1715, a period of seventy years, virtually no sunspots were visible despite the best efforts of astronomers to find them. This strange phenomenon, the "Maunder Minimum," poses yet another puzzle for solar physicists to solve.

• • • • •

But let's get back to Hale. After discovering that sunspots possess powerful magnetic fields, Hale embarked on a systematic study of their behavior through a full solar cycle. He found several important regularities.

First, sunspots appear at the surface in pairs (or "bipoles") that have opposite magnetic polarities. In the Northern Hemisphere, all the "leading" spots—those ahead in the direction of solar rotation—have the same polarity. In the Southern hemisphere, the leading spots have the polarity opposite to the leaders in the Northern Hemisphere (fig. 10.4). Finally, the polarities of the leaders (and of course the followers as well) reverse in the next solar cycle. So, although the number of spots varies in an eleven-year cycle, their magnetic polarities reverse in a twenty-two-year cycle.

These empirical rules, coupled with Carrington's discovery of the latitudinal variation of new spots and Schwabe's estimate of an eleven-year cycle, cry out for a physical explanation. What is going on under the surface of the Sun to create

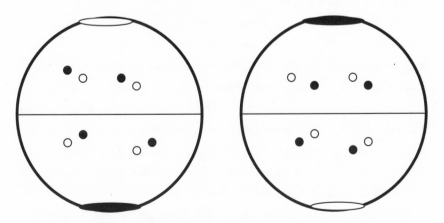

FIG. 10.4 Hale's rules governing the polarities of sunspots in two successive eleven-year cycles. A complete cycle takes twenty-two years.

this remarkable order? Theorists have been pondering this question ever since Hale made his stunning discovery. They have been guided by the insights of a skilled nontheorist.

A QUALITATIVE EXPLANATION

Horace Babcock, a solar astronomer at Mount Wilson Observatory, followed in Hale's footsteps. He was also a talented instrumentalist who invented the magnetograph, an electronic device for measuring weak magnetic fields. With this instrument he discovered that large areas of a weaker field (the so-called active regions) surround the sunspots and gradually spread out to cover much of the surface. Moreover, he noted that the geographic poles of the Sun are also magnetic poles, similar to those of the Earth, and that the polarity of each pole reverses just before the number of sunspots reaches its maximum in a cycle. Oddly enough, the north and south poles don't always reverse polarity simultaneously, as one might expect if a direct connection existed between them. These important additions to the Hale rules gave theorists much more to think about.

In 1961, Babcock published a conceptual scheme for the origin of the solar cycle that has been enormously influential (fig. 10.5). To follow his scheme, we need to recall that magnetic fields can be imagined as elastic field lines that can be stretched indefinitely without breaking. Second, the gas in the Sun's interior is actually a "plasma," composed mainly of free electrons and protons. These charged particles lock onto a magnetic field by spiraling around the field lines. In this way, the field and the plasma are forced to move together. (Astronomers speak of the field as "frozen" into the plasma.) Below the surface of the Sun, the field is relatively too weak to resist the convective motions of the plasma. It is constrained to follow, and stretches if it has to. In the corona, however, the magnetic field is relatively strong compared to the plasma motions and therefore forces the plasma to follow the field lines. Finally, if two oppositely directed fields touch, they slowly cancel and convert their energy to heat.

Babcock incorporated the key concept of frozen magnetic fields in his scheme. In his picture (fig. 10.5) the cycle starts with a weak field connecting the Sun's poles. Beneath the surface, the latitude variation of rotation wraps the field lines

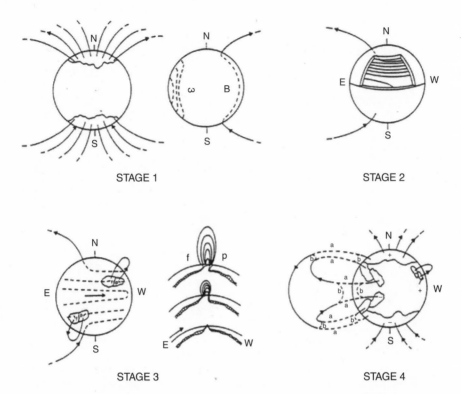

FIG. 1 0. 5   Horace Babcock's scenario for the generation of magnetic fields and sunspots by the action of differential rotation.

around the Sun in a spiral pattern of magnetic "ropes." As the wrapping proceeds, the number of adjacent field lines increases, which corresponds to an increase in magnetic field strength of the ropes.

After about three years, the magnetic pressure in the ropes is supposed to reach a critical value, sufficient to expel most of the gas within them. As a result, the ropes become buoyant. Note that, because of the way the field is wrapped, this transition occurs first at high latitudes. At random points along a high-latitude rope, a kink or loop appears and rises through the surface. The feet of such a loop become a pair of visible sunspots.

All the leading spots in a hemisphere will have the same polarity because of the alignment of the ropes, and the leaders in opposite hemispheres will have op-

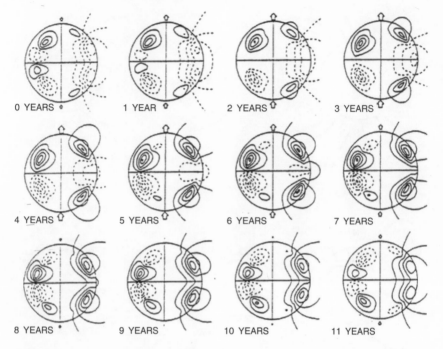

FIG. 10.6 In this kinematic model of the sunspot cycle, toroidal fields (shown on the left hemisphere of each sphere) migrate to the equator. The poloidal fields (shown in the right hemispheres) break through the surface and appear as azimuthal bands of sunspots. The strength of positive fields is drawn in solid contours, negative fields in dashed lines.

posite magnetic polarities. As the wrapping progresses, the field reaches the critical value at lower and lower latitudes. This result explains why new spots occur at lower latitudes, the reason for the "butterfly diagram."

Babcock accounted for the reversal of polarity of the poles in the following way. Because of the way the ropes are inclined to the equator, a pair of spots will emerge with the follower closer to the nearest pole. Moreover, during the rise of the cycle, the followers have the polarity opposite to that of the nearest pole. Therefore, as the field of the spots spreads over the surface, the follower's field will reach the nearest pole before the leader's, will cancel the pole's weak magnetic field, and will replace it with its own. At the end of the cycle, all spot fields will have merged and canceled, leaving a simple dipolar field that connects the north and south poles of the Sun, but in the reversed direction. A new cycle is ready to begin.

Robert Leighton, the Caltech professor we met in chapter 1, improved on Babcock's scheme and made it somewhat more quantitative. He had discovered that the surface of the Sun is covered with "supergranules," which are convection cells approximately twenty times larger than the ordinary granules, having lifetimes of about a day. The cells spread out horizontally with speeds of a fraction of a kilometer per second when they reach the surface. Leighton suggested that these horizontal flows could spread magnetic fields away from the spots, over the surface and up to the poles. He calculated how such cells could shuffle the feet of magnetic loops over the surface in a kind of random walk (note 10.1). With this mechanism he was able to reproduce the observed patterns of large-scale magnetic fields.

Babcock's scheme, modified by Leighton, is consistent with all of the Hale rules and gives a qualitative explanation for the origin of the magnetic cycle. This is no mean feat. Although this model lacks a quantitative physical explanation for many of the steps, its basic idea of discrete magnetic ropes that wrap around the sun has guided theorists for almost forty years.

TOY SUNS

As ever more powerful computers became available in the 1970s and 1980s, theorists began to produce numerical simulations of the solar cycle. This was a time when the rotation of the Sun's interior was still uncertain, so a theorist was free to guess it. He could then construct a time-dependent "kinematic" model of the Sun, in which the gas motions were specified in advance. For each moment of time, he would have to solve a particular, nasty mathematical equation in three spatial dimensions and time. Then, starting with a simple pole-to-pole field in analogy to Babcock's model, he could follow the evolution of the field.

To simplify the calculation, theorists usually assume that magnetic fields are always symmetrical about the rotation axis, meaning there are no variations with longitude but only with latitude and depth. In effect, such a simulation averages over all the sunspots within a fixed band of latitude. With this compromise, the Hale rules can be reproduced to some extent.

In order to proceed, some assumptions had to be made concerning the variation of rotation speeds in depth and latitude. In the early days, theorists normally assumed that the angular speed is constant on *cylinders* and decreases out from the rotation axis (see fig. 5.2). That would accomplish two things: it would account for the pole-to-equator variation of rotation speed seen at the surface, and it would also produce the correct migration of sunspots toward the equator during the cycle. Note that each cylinder in figure 5.2 rubs against its neighbors, creating a shearing flow throughout the convection zone.

The simulation starts with an initial pole-to-pole ("poloidal") field, just as in Babcock's scheme. The assumed shearing flow wraps the field lines around the Sun to form "toroidal" ropes and amplifies them. This is the "omega effect." In addition to rotation, rising and sinking convective motions are introduced in the simulation. Eventually these convective motions punch kinks in the toroidal field and carry them upward.

And now another new factor appears. As a convective cell rises and spreads horizontally, carrying the field with it, the Coriolis force twists the kink—clockwise in the Northern Hemisphere and counterclockwise in the Southern (see note 8.1 on the Coriolis force). This magical effect, the so-called alpha effect, was discovered (or invented) by Eugene Parker, the famous astrophysicist at the University of Chicago. The alpha effect converts east-west fields to north-south fields, which migrate to the poles and reverse their polarity, allowing a new cycle to begin, as before.

In figure 10.6 we see the results of a typical simulation, performed by Michael Stix, a German theorist, in 1976. On the right-hand side of each disk are the poloidal (north-south) components of the field; on the left side, the toroidal (east-west) components. At three years into the cycle, a twisted rope in each hemisphere breaks through the surface as the analogue of a new belt of sunspots. As the cycle advances, these belts expand into the corona and migrate toward the equator. After ten years, the belts merge and cancel, leaving behind a weak reversed poloidal field to start the next cycle.

Pretty impressive, isn't it? When a model like this begins to approximate reality, we begin to think we understand the physics behind the cycle. Unfortunately, the model turns out to have a fatal flaw. In order to get the new fields to migrate toward the equator, Stix had to assume that the angular speed of rotation in the convection zone *decreases* with increasing depth. In 1989, Tim Brown and Cherilynn Morrow killed that idea. Using the Fourier tachometer at the Sacramento Peak Observatory, they discovered that the rotation speed in the convection zone varies in latitude, but *not* with depth.

Stix was not alone in this predicament. All kinematic models of the solar dynamo suffer from the same inadequacies. And as we have seen earlier, more complete *dynamic* models that incorporated much more physics of solar convection still failed to reproduce the observed pattern of the Sun's internal rotation, much less Hale's rules of sunspot polarities. Moreover, no model was able to predict the most basic property of the cycle, its period. Only by tuning the parameters of a model was it possible to obtain the correct eleven- (or twenty-two) year period.

PROBLEMS, PROBLEMS

As theorists continued to grapple with the task of predicting how the Sun produces its beautiful magnetic regularities, they recognized several other problems. Gene Parker had shown as early as 1975 that a magnetic rope would rise through the convection zone much too quickly. It would reach the surface in a few months, long before it could acquire sufficient field strength (say 3000 gauss) to produce a sunspot. A possible cure was to assume that the amplification of the field occurs very deep in the convection zone, ensuring a much longer rise time, with more time for amplification.

Recent studies have raised problems with that idea, however. If a rope has field strength less than about 60,000 gauss, it will be too flexible. The Coriolis force will twist it too much as it rises, and it will also erupt at too high a latitude. If, on the other hand, a rope has field strength larger than 160,000 gauss, it will emerge in the sunspot latitudes, but it will be too stiff. The twists will be too small and wouldn't produce enough poloidal field. So only a field of about 100,000 gauss at the base of the convection zone can produce the desired results. But such a strong field can't be stored for several years, even at the very bottom of the convection zone.

The theorists were between a rock and a hard place. Helioseismology showed the way out of this dilemma.

A SOLAR CONVEYOR BELT

In 1989, Tim Brown and Cherilynn Morrow discovered that the pattern of rotation changes rather abruptly near the base of the convection zone (see a later map, fig. 9.1). In the convection zone, the angular speed of rotation varies with latitude but not with depth. In the radiative zone the angular speed doesn't vary with *either* latitude or depth, but it rotates as a rigid body would, at the speed the convection zone has at latitude 30 degrees. Therefore, where the two regions meet, the convection zone rotates *slower* near the poles than the radiative zone, and *faster* near the equator. At these latitudes the two regions *rub against each other* in a sliding or shearing motion. That result would prove to hold the key to the solution.

Peter Gilman, Ed DeLuca, and Cheri Morrow pointed the way. Gilman and DeLuca are both experienced hydrodynamicists, and they immediately grasped the importance of Morrow and Brown's discovery. These three offered a qualitative scheme to account for the differential rotation seen throughout the convection zone (fig. 10.7). To account for the faster rotation at the equator, angular momentum (spin) must be transported somehow from pole to equator. To balance the books, angular momentum must also be fed back to the poles, presumably in the transition layer at the base of the convection zone.

How might this occur? Very likely through the rubbing of one zone on the other, in a thin transition layer at the base of the convection zone. At high latitudes, the transition layer is rotating faster, and so it pumps angular momentum (spin) into the convection zone. This momentum is transported, somehow, to low latitudes, where it leaks back into the transition layer. The same quantity of angular momentum is then transported poleward, somehow, in the layer. Round and round, this process proceeds in a cycle, driven presumably by the forces of convection.

It is an appealing picture, and the authors quoted some support from other observations. But the devil is in the details. They were as yet unable to specify

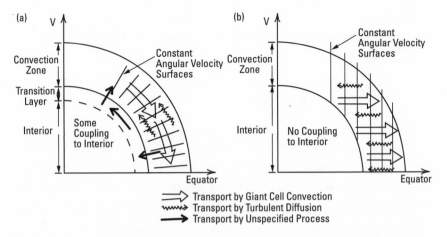

FIG. 10.7 (a) A proposal for producing the observed surfaces of constant rotation. Angular momentum is transported toward the equator by giant cells in the convection zone and back to the poles by some unspecified process through a thin transition layer. Turbulent diffusion in the convection zone acts as a drag on the transport (wiggly arrows). A dynamic model of convection, based on giant cells, yielded the unrealistic pattern of rotation shown in (b).

what kind of frictional force couples the transition layer to the convection zone. Nor could they propose a physical mechanism for the poleward transport of momentum in the transition layer.

The transport mechanism in the convection zone was equally mysterious. Previously, Gilman had proposed that giant convection cells would transport angular momentum to the equator, but his extensive calculations showed otherwise. As the right panel in figure 10.7 shows, giant cells transport momentum away from the rotation axis, not toward the equator. Nevertheless, Gilman and DeLuca were able to sketch how a kinematic dynamo might work in a thin transition layer, now that its internal rotation was known.

A basic problem arose immediately. The model predicted that, below a middle latitude, toroidal ropes would migrate to the equator, in agreement with the butterfly diagram (see fig. 10.3). But the model also predicted that above that latitude, the ropes would migrate to the *poles*, in contradiction to sunspot observations.

Clearly, some essential idea was missing. Despite these limitations, this research focused attention on the transition layer as the likely site of the solar dynamo.

THE TACHOCLINE

In 1992, Ed Spiegel of Columbia University and Jean-Paul Zahn of the Nice Observatory gave the name "tachocline" to this transition layer and were the first to investigate its dynamics. They proposed a way that the tachocline could transport angular momentum to the poles, one of the essential links in the cycle we just described. Basically, they claimed that the shearing motions would generate waves on the tachocline that would carry angular momentum to the poles.

Paul Charbonneau and friends shot down that idea in 1999. The tachocline is stable, they claimed. Hydrodynamic waves can't develop on its surface. However, Peter Gilman and Peter Fox showed that the presence of even a weak toroidal magnetic field (1000 gauss) could destabilize the tachocline. The shearing flow would then generate *magnetic* waves that carry angular momentum to the poles. So here is evidence that the existence of differential rotation may depend on magnetic fields that depend, in turn, on the differential rotation. The two basic effects might be coupled, like two kids on a seesaw. This is still a highly controversial subject, however. The physical processes involved in generating differential rotation are still being debated.

As helioseismic observations improved, the depth and thickness of the tachocline could be determined more precisely. In 1996, Alexander Kosovichev at Stanford University analyzed two years of oscillation data obtained by Martin Woodard and Ken Libbrecht at the Big Bear Solar Observatory. He determined that the layer is about one-tenth of a radius thick and its center lies at a fractional radius of 0.692. Later, Jesper Schou used SOHO data and Paul Charbonneau used LOWL data to refine these results. The latest estimates say the tachocline has a thickness of 0.039 radii (or about 28,000 km) and is centered at a radius of 0.693, near the base of the convection zone at 0.713R.

Figure 10.8 shows how the tachocline straddles the precise boundary between the neighboring zones. In this thin layer the rotation speed decreases along an

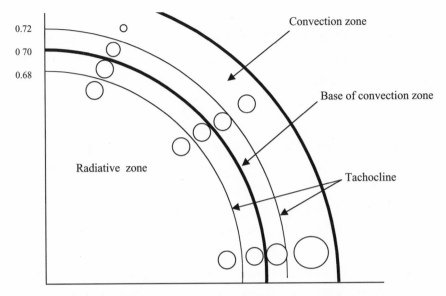

FIG. 10.8   The tachocline lies between the convection and radiative zones. At high latitudes, the rotation speed (shown by the size of the circles) decreases outward from the radiative zone. At low latitudes the speed increases outward. Therefore, the radiative and convection zones rub against each other at high and low latitudes.

outward radius at high latitudes and increases outward at low latitudes. The rotation speed increases from pole to equator everywhere in the tachocline and convection zone.

A NEW KIND OF DYNAMO

Since the early 1990s, solar physicists have focused on the tachocline as the seat of the Sun's dynamo, for two main reasons. First, the helioseismic observations tell us that the tachocline is the *only* place in the Sun that a radial variation of rotation speed, necessary for amplifying toroidal fields, exists (but see note 10.2). Second, a magnetic rope could be stored indefinitely in the radiative zone just below the tachocline, because any tendency to rise buoyantly from there would be suppressed (note 10.3). A rope could be held there long enough to reach the critical field strength of about 100,000 gauss.

Now, it's one thing to say, "Here lies the dynamo," and another to demonstrate how such a dynamo might work. Many types of dynamos have been proposed during the past thirty years. Some were kinematic, others were dynamic, but all of them were incomplete in some way or other. Now with the tachocline in view, the question was, "How can a dynamo work there?" Gene Parker, whom we have met many times before, came up with an interesting possibility in 1993. He built on earlier work by such theorists as Alexander Ruzmaikin.

In the conventional scheme, the alpha effect, which produces a poloidal field from a toroidal field, and the stretching (or "omega" effect), which produces a toroidal field from a poloidal field, are both located in the convection zone. In Parker's "interface dynamo," the two effects are separated by a sharp boundary (analogous to the tachocline) between the radiative and convective zones. Convection cells above the interface would produce the alpha effect, while a radial shear in rotation speeds at the interface would produce the omega effect (for details read note 10.4).

The key idea here is that magnetic ropes would be confined below the interface, in the radiative zone, where they are not buoyant (see note 10.3). A radial shear in the speed of rotation below the interface would amplify these ropes until their magnetic pressure reached the surrounding gas pressure. They could be as strong as 100,000 gauss and still not pop up to the surface. Meanwhile, shreds of the upper part of a rope would continuously leak across the interface, injecting a weaker field into the convection zone, where the usual alpha effect of rising cells would generate a poloidal field.

In order for this scheme to work, the field has to diffuse through the gas very slowly below the interface and very rapidly above it. Parker showed that if rates of diffusion differed by a large factor, say a hundred, the scheme works nicely. There is good reason to think he could be right, because turbulence in the convection zone would enhance diffusion enormously, while the strong field in the radiative zone would suppress turbulence and, therefore, diffusion.

The net effect of all this machinery would be to create a toroidal field above the interface that steadily increases in strength. Parker showed that the field takes the form of a toroidal "dynamo wave" that drifts slowly in one direction. (We can see an example of such a wave in Stix's model, fig. 10.6, where the toroidal field in the left-hand quadrant drifts from pole to equator in eleven years.)

Parker demonstrated a simple example of an interface dynamo that generates a traveling dynamo wave. The next step would be to incorporate such a dynamo in a more realistic model of the Sun.

A GLOBAL INTERFACE MODEL

Parker, in his usual way, laid out the key elements in an interface dynamo and left the messy details to others. Paul Charbonneau and Keith MacGregor, the two bright young theorists at the High Altitude Observatory, picked up the interface idea in 1997 and ran with it. They constructed numerical models of the solar cycle, replacing Parker's infinitely thin interface with a slim tachocline. They wanted to see whether they could reproduce Hale's rules of magnetic polarities by calculating only the mean field at each latitude. This would obviously be a lot

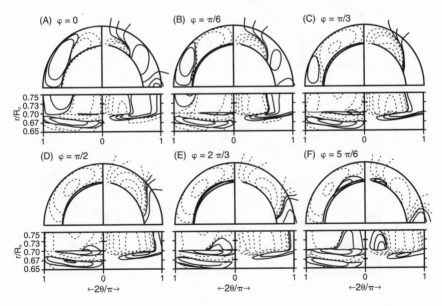

FIG. 10.9 A kinematic model of the solar cycle, based on Parker's interface dynamo. See note 10.5 for a description.

easier than trying to reproduce individual spot groups. Their basic model has a spherical shape and, for simplicity, the fields are assumed to be symmetric about the rotation axis. Also, all the motions are prescribed in advance, so it is a kinematic model with, however, a realistic rotation profile.

Parker considered only the effects of a radial variation of rotation speed. By including both radial and latitudinal variations, as seen in the real Sun, Charbonneau and MacGregor found more complicated dynamos. One type was driven by purely latitudinal variations, a second type was driven by purely radial variations. Both types could coexist and in certain circumstances could interfere destructively.

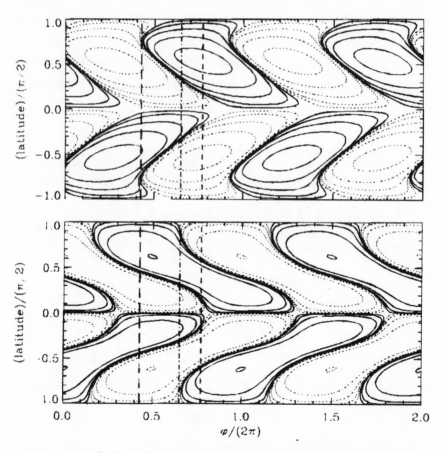

FIG. 10.10 The butterfly diagram predicted by the model shown in figure 10.9.

Figure 10.9 illustrates the progress of a particular dynamo model through an eleven-year cycle. The figure is similar to Stix's plot (fig. 10.6) in the way it plots poloidal and toroidal fields. (If you are interested in its details, read note 10.5.) You can easily follow the migration of both types of fields toward the equator and a reversal of the magnetic field at the poles of the sun.

In figure 10.10 we see the predicted "butterfly diagrams" of a toroidal field under the interface (upper panel) and a radial field at the surface (lower panel). Compare the upper panel with figure 10.3, the real butterfly diagram, and you will agree that the model does a pretty good job. It illustrates how Parker's idea of an interface dynamo can work reasonably well.

Like all kinematic models of the solar cycle, this one is incomplete in many ways. For example, it can't predict the period of the cycle or the strength of the magnetic fields. Those tasks remain for nonlinear dynamic models that incorporate more of the physics of convection. As we have seen, however, such models are still some distance from predicting the observed profile of rotation. Predicting how rotation, convection, and magnetic fields interact is an even greater challenge. But we are getting closer to a solution.

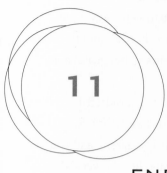

11

AD ASTRA PER
ASPERA — "TO THE
STARS THROUGH
ENDEAVOR"

IF THE SUN OSCILLATES, why shouldn't other stars? If we can learn so much about the inside of the Sun by studying its vibrations, why can't we do the same with other stars? Why not, indeed?

Astronomers have known for over a hundred years that many stars vary in brightness with perfectly stable periods. Some of these are actually binary systems, which vary in brightness as the two stars orbit each other. However, the more interesting and more valuable variables are the pulsating stars. Among them are the Cepheids, giant yellow stars that are intrinsically much brighter than the Sun. John Goodricke, a deaf-mute English amateur astronomer, discovered the prototype, Delta Cephei, in 1784.

The true nature of these stars remained hidden until 1912. At that time, Henrietta Leavitt, a demure assistant to the famous Harvard astronomer Harlow Shapley, was cataloging variable stars in the Large Magellanic Cloud, a satellite galaxy of our own Milky Way. She discovered thousands of variable stars, among them the Cepheids, with their characteristic shark-fin light curve (fig. 11.1). She noticed that their periods varied from one to fifty days and that the fainter they were, the shorter their periods were. In fact, when she plotted brightness against period she discovered a nice linear relationship. Because all these Cepheids were at essentially the same distance from Earth, in the Large Magellanic Cloud, she

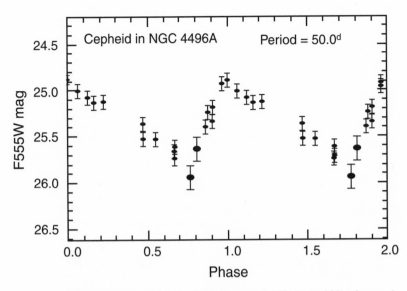

FIG. 11.1   The light curve of this Cepheid, discovered with the Hubble telescope in the distant galaxy NGC 4496A, has a typical shark-fin profile.

recognized that the period of a Cepheid is a reliable measure of its intrinsic brightness. This discovery was of tremendous importance. It meant the Cepheids are "standard candles" that can be used to measure intergalactic distances. Shapley determined the intrinsic brightness of a Cepheid whose distance was known, and from then on the Cepheids became an invaluable tool for mapping the universe.

Sir Arthur Eddington, the dean of British astrophysicists, was intrigued with the Cepheids. In 1917 he proposed a pulsation model for the star based on simple physical arguments. He suggested that a Cepheid expands and contracts radially, like a balloon, growing brighter as it expands. Its pulsation, he argued, is caused by a competition between an overpressure, which causes the star to expand, and gravity, which causes the star to contract. Unfortunately, his explanation turned out to be wrong in its details. The star reaches maximum brightness when it is expanding most rapidly, not at maximum size. This observation contradicts his model, but his basic idea that the Cepheids pulsate blazed the trail for others to follow. (See note 11.1 for current explanations.)

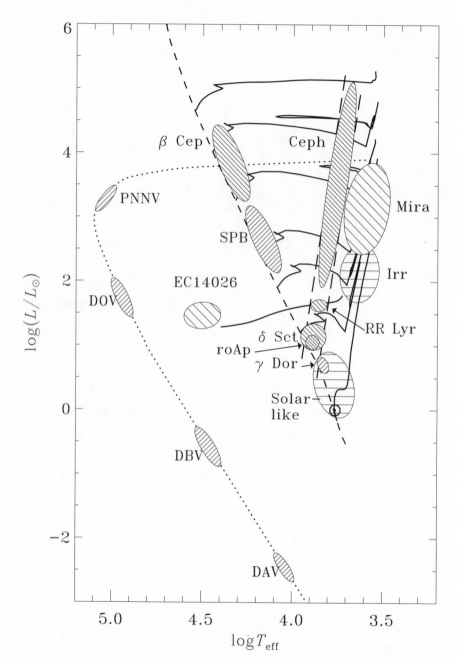

FIG. 11.2 A collection of variable stars, sorted according to their colors or surface temperature and absolute brightness.

A simple radial pulsation with one period does not provide much information about the interior of a star, however. The five-minute oscillations of the Sun are so informative because they are nonradial and hence have a huge spectrum of periods and many modes of oscillation.

Stellar astronomers were fully aware of the scientific potential of stellar vibrations as early as 1970. Since then they have pursued their quarry relentlessly, in parallel with the helioseismologists. They have discovered several classes of stars that vibrate nonradially and have been able to draw some definite conclusions about their interiors. In this chapter we'll explore some of their discoveries in the new science of asteroseismology.

Bear in mind that this is a difficult game. The stars are much fainter than the Sun, which limits the accuracy of measurements. Even worse, we can't resolve the surfaces of stars in order to sort out different patterns of vibrations, as we can with the Sun. And as we will see, the rapid rotation of some interesting stars greatly complicates any interpretation of their pulsations. Nevertheless, astronomers have made progress.

A BRIEF PRIMER

A tremendous variety of stars pulsate. There are red giants, white dwarfs, yellow sunlike stars, blue supergiants, and many more. Figure 11.2 is an attempt to sort out this morass. Here the stars are plotted according to two basic characteristics: their intrinsic luminosity (relative to the Sun), and the "effective" temperature of their surfaces. (This diagram is a portion of the famous Hertzsprung-Russell diagram; see note 11.2.) The curves are theoretical evolutionary tracks that stars of different masses follow as they age. Note the so-called instability strip, shown as two dashed lines sloping downward on the upper right. Every star with a mass less than eight solar masses passes through this strip during its lifetime and becomes unstable. Three important classes of pulsating stars lie here: the Cepheids, the RR Lyrae variables, and the Delta Scuti variables. (See note 11.3 on star names.)

Some Cepheids, as you can see, are 100 to 100,000 times as bright as the Sun. A Cepheid can vary in brightness during its cycle by a factor of 1.5 to 6, a huge change compared to the Sun's minuscule 0.1%, making them relatively easy to spot. Most of these Cepheids have periods between one and fifty days. In contrast, the short-period Cepheids or RR Lyrae stars pulsate with periods of less than a day.

Delta Scuti stars (see fig. 11.2) are giant white stars, about three times as bright as the Sun. They pulsate with periods between half an hour to eight hours, and vary in brightness by less than a factor of two. Way down in the diagram lie the sunlike stars, like the bright stars Procyon and Alpha Centaurus. Judging from what we know about the Sun, we should certainly expect these to oscillate.

Even farther down the figure lie the white dwarfs, which all have the letter D as a prefix to their names. Friedrich Bessel discovered the first white dwarf in 1844. He noticed that Sirius, the brightest star in the sky, oscillates back and forth along the line of sight, which suggested that it had an unseen companion. In 1863 Alvin Clark, an optician, was able to resolve the tiny star with his superb telescope. Although white dwarfs are the faintest of pulsating stars, and therefore the most difficult to study, they have proved to be among the most productive stars of all. So let's begin with them.

SMALL IS BEAUTIFUL

White dwarfs are truly remarkable objects. Theorists tell us that as an average star such as the Sun ages, it consumes most of the hydrogen in its core, and later most of the helium, which it converts to carbon. As the core evolves, the star's atmosphere expands and cools. The star becomes a luminous red giant, with an atmosphere the diameter of Earth's orbit. Eventually the star ejects a fraction of its mass and creates a so-called planetary nebula (labeled PN in fig. 11.2), a bright shell of glowing hydrogen gas. At the center of the nebula lies an Earth-sized star, with about one solar mass, which will slowly evolve into a white dwarf with about half the mass of the Sun. With no nuclear fuel to burn, the dwarf simply cools and shrinks over billions of years. The coolest dwarfs have estimated ages of 9 to 10 billion years and set lower limits to the age of our galaxy. In 2002 one was found with an estimated age of 13 billion years, in satisfactory agreement with other estimates of the age of the universe.

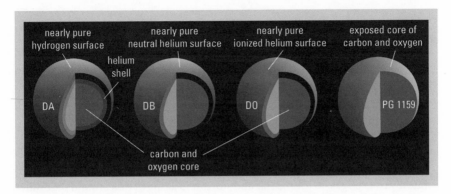

FIG. 11.3   White dwarfs differ according to their surface chemical composition and surface temperature.

A white dwarf's surface temperature is much higher than that of the Sun, anywhere from 10,000 to over 100,000 K, and for that reason the dwarf is called "white." Its intense gravity, ten thousand times that of the Sun's, crushes its atoms and forces them into a strange state, a "degenerate" electron gas (see note 11.4). A cubic centimeter of the stuff in its interior could weigh ten tons on Earth.

White dwarfs fall into several distinct categories that differ according to the nature of their surface layers and spectra. As we can see in figure 11.3, the DAV and DBV dwarfs have very thin surface layers of hydrogen and helium, respectively, and surface temperatures of about 12,000 K and 28,000 K. The DOV dwarfs are extremely hot, with surface temperatures up to 150,000 K. The coolest dwarfs, the DCs, emit no spectral lines at all, but only a smooth spectrum. All four types have cores composed of carbon and oxygen.

No one knows yet why the dwarfs divide into these spectral groups. Evry Schatzman, a French astrophysicist, proposed in 1949 that a dwarf's extreme gravity would cause all the heavy elements to settle into the interior, leaving the surface layers to hydrogen or helium. But the details are still uncertain.

Incidentally, many white dwarfs have enormous magnetic fields. The star REJ037-853 has one of the strongest fields, 340 million gauss, or about one hundred thousand times that of a sunspot.

DWARF NOVAE

If a white dwarf happens to orbit another star in a binary system, interesting things can happen. For example, take the binary U Geminorum, which has a normal red star and a white dwarf revolving practically nose to nose. Every few weeks, the system flares up in brightness by factors of 6 to 100. Such violent outbursts, called "dwarf novae," have also been seen in X rays. Evidently, the powerful gravity of the white dwarf tears gas off the red star and stores it in a ring or disk around the dwarf. Occasionally, a surge of in-falling mass creates a hot spot in the disk or on the dwarf's surface. The hot spot emits a tremendous burst of flickering light.

In the early 1970s Edward Nather and Edward Robinson at the McDonald Observatory in Texas, together with Brian Warner and John McGraw at the Cape Observatory in South Africa, began to use high-speed photometers to examine the flickering light from such dwarf novas. They learned that the light doesn't vary erratically, but periodically, with multiple periods between 100 and 1000 seconds.

In 1972 Warner and coworkers examined the remnants of a "classical" nova that had exploded in 1934. This old nova, DQ Hercules, is an *eclipsing* binary, composed of a white dwarf and a red dwarf, which makes it very special. (Red dwarfs are faint cool stars, with about half of the Sun's mass.) The orbit of the white dwarf and its companion is oriented so that each star eclipses the other in a period of 4.6 hours. Such binaries are especially valuable because they can yield the stars' masses and radii. In addition the light from this binary was varying in a pure sinusoid, with a period of 71.06550 seconds. This period is much longer than the radial pulsation period of a white dwarf, which would be a few seconds. Warner and friends concluded that the dwarf was vibrating nonradially and attributed the oscillations to trapped gravity waves.

The theory of stellar oscillations had been developed over many years. Back in 1955, Paul Ledoux in Belgium and Evry Schatzman in France laid out the theory of nonradial oscillations and pointed to the possibility of gravity modes. When Warner and his coworker Edward Robinson announced their startling conclusions, nonradial oscillations suddenly became interesting again. Yoji Osaki and Carl Hansen, and Arthur Brickhill, refined the basic calculations and applied them to white dwarfs. These calculations suggested that an isolated white dwarf, with no orbiting companion, might vibrate independently. Arlo Landolt, an astronomer at Louisiana State University, discovered one in 1968.

In the early 1970s, Warner, Robinson, Nather, and John McGraw began an intensive survey of white dwarfs, looking for other examples of vibrating white dwarfs. In a crucial investigation, McGraw and Robinson used a normal star as a reference to eliminate the twinkling of a dwarf's light that is caused by the Earth's atmosphere. They discovered that most isolated white dwarfs *don't* oscillate, they only twinkle. But by 1976 they had compiled a list of ten genuine examples, some with *several dozen* periods running simultaneously. The periods fell in the range of 100 to 1000 seconds. Some periods were absolutely stable, but some dwarfs could switch periods in a few hours. The amplitudes of the oscillations could also change, sometimes periodically, sometimes at random.

Their preferred explanation for these oscillations was nonradial gravity modes, because only these have periods in the observed range. (Pressure modes, like those observed in the Sun, have much shorter periods.) They could usually find predicted periods that matched their observations quite well. But this explanation had its difficulties. The theory predicted many more modes than they observed, and the modes they did see were sometimes high harmonics. For example, one dwarf was vibrating in four modes with $L = 2$, but with $N = 6, 7, 8$, and 9. Where were the modes from $N = 1$ to 5?

On the other hand, the Texas astronomers could provide a very satisfying explanation for another DAV white dwarf, R548. It is the prototype for a whole class of dwarf pulsators, the so-called ZZ Ceti stars. R548 has a complicated light curve, which varies smoothly over a period of about one and a half days. Analyzing the curve, the astronomers discovered that R548 pulsates in four modes simultaneously, with periods of 212.77, 213.13, 274.25, and 274.77 seconds. The periods were stable over at least nine years, but the light curve varies because of the "beats" among the different pulsation periods.

WHY DO THE DWARFS OSCILLATE?

In each type of dwarf, a thin region of partial ionization of the most abundant element lies just below the surface (see fig. 11.3). Thus, in the DAV stars, some of

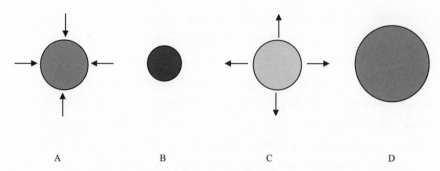

FIG. 11.4   The kappa mechanism of variable star pulsation. See text for explanation.

the hydrogen atoms have lost their electron; in the DBV, some of the helium atoms have lost an electron. In the DOV stars, carbon and oxygen are partially ionized. These ionization zones can store (and later release) vast quantities of energy, because stripping an atom of an electron requires a relatively large energy, 13.6 electron volts for hydrogen, 25 volts for helium. This energy can be released as a photon when an atom picks up a free electron.

The cause of the pulsation is basically the same as in the Cepheids, the so-called kappa mechanism. (Kappa is the Greek letter assigned to the opacity in a stellar atmosphere.) Under the surface of a DBV dwarf, for example, lies a layer of partially ionized helium. When this layer is compressed (fig. 11.4A), its temperature rises, causing the gas to become more opaque to radiation. As a result, radiant energy flowing from the interior is damned up and stored as the energy of helium ionization. The layer continues to shrink to a minimum diameter (fig. 11.4B). Eventually, the increased pressure causes the layer to expand and cool, which in turn reduces the opacity. The stored energy is converted back to radiation and floods out as the star expands. The star is brightest when it is expanding most rapidly (fig. 11.4C). Finally, gravity causes the layer to stop expanding and start collapsing (fig. 11.4D), and so the cycle repeats itself. In this way, an ionization zone acts as a gate for radiant energy, storing and releasing it periodically, and producing luminosity variations.

But what about the gravity modes? These are waves that are excited by the expansion and contraction of the star's envelope and are *trapped* in the thin ionization zone. If the vertical wavelength of a mode happens to fit exactly in the thickness of the zone, the wave will be trapped as a nonradial gravity mode. In such a standing wave, buoyancy and gravity combine to produce the wave's oscillation.

WHAT CAN WE LEARN?

As expected, when one has good observations of a large number of modes, one can derive all sorts of interesting information about the structure of a white dwarf. One of the best early examples was investigated by John McGraw, Donald Winget, and coworkers in the early 1980s. The star in question has the unlovely moniker of PG1159-035. It is a "pre–white dwarf," a very hot planetary nebula nucleus that will evolve further to become a typical helium-rich DOV white dwarf (consult fig. 11.2). This star has only about half the mass of the Sun but radiates about one hundred times as much energy.

At least eight pulsation periods were seen in its light curve: 390, 424, 450, 495, 516, 539, 645, and 832 seconds. These periods are too long to be anything other than nonradial gravity modes. They have been identified as modes with $L = 1$ to 3 and, surprisingly, with radial number N higher than about 25.

It turns out that the average differences in these periods are all integral multiples of 8.81 seconds; that is, the periods are spaced uniformly. Steven Kawaler, a professor now at Iowa State University, showed how such behavior depends on the mass of the star. Consulting his models, he derived a mass of 0.6 solar masses, in good agreement with other estimates for white dwarfs.

A closer look at the periods, including two new ones, shows some small variations from uniform spacing. Kawaler predicted that these variations occur when the vertical wavelength of a mode fits nicely into the thickness of a surface layer, the "trapping of modes." From this correspondence he was able to derive layer thicknesses, some as small as 50 km.

All this is very well, but a nagging question remains. Where are the gravity modes with N less than 25? The kappa mechanism is the first choice for generating the modes, but it would excite many more modes than are seen. Kawaler proposed that some unknown form of filter suppresses the low-N modes, but the lack of a physical explanation casts some doubt on the kappa mechanism.

Another possibility, at least for these very hot DOV stars, is the nuclear burning of carbon and oxygen in a shell surrounding the deep core. Theoretical models show that the temperature in the shell and hence the burning rate can oscillate and pump energy into gravity modes. This is the "epsilon mechanism," but it too has a flaw. It predicts periods three to five times shorter than are observed in DOV stars. That might suggest that something is missing in the standard models of the interior of such stars. The situation here was therefore similar to that facing helioseismologists in the 1980s, namely, how to modify the model of the star's interior to match the observed pulsation frequencies.

NETWORKING, FOR A SHARPER VIEW

In the late 1980s, asteroseismologists realized that the only way to make further progress, especially on the tantalizing white dwarfs, was to resolve more periods in the oscillation spectrum. Just like the solar astronomers, they were driven to make longer and more continuous observations of stellar pulsations. In particular, they had to eliminate the daytime interruptions of their time series. Just like the solar astronomers, they argued for a network of cooperating observatories, spaced around the globe, that could observe a chosen star for as long as possible during several nights.

Edward Nather and Donald Winget of the University of Texas were two of the most ardent proponents of this scheme, and they were motivated by the wealth of theoretical predictions made by Steven Kawaler. Their efforts bore fruit in 1989, when they persuaded nine observatories to band together as the Whole Earth Telescope, or WET. The participants included two-meter-class instruments in Texas, France, Brazil, Chile, and the Canary Islands. All were equipped with high-speed photometers to measure the complicated variations of a star's light. Some of these instruments had a second channel, which was used to observe a nonpulsating star simultaneously with the target star, thus eliminating the twinkling of the Earth's atmosphere.

The first target was the pre–white dwarf with the ugly name and beautiful pulsation spectrum, PG 1159-035. Continuous observations extending over 264 hours were obtained in March 1989. The observations from each station were e-

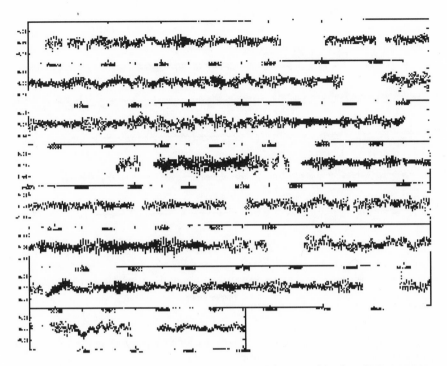

FIG. 11.5   Whole Earth Telescope observations of the pre–white dwarf PG 1159-035 reveal a very complicated light curve of brightness versus time.

mailed to the center at the McDonald Observatory in Texas, where they were reduced and analyzed during the observing run.

An analysis of its light curve (fig. 11.5) yielded no fewer than 125 periods, ranging from 385 to 1000 seconds (fig. 11.6). Of these, 101 could be unambiguously identified with gravity modes of L = 1 and 2 and most of the corresponding values of the azimuthal number, M. At last, there was no longer any doubt that the pulsations were indeed caused by standing gravity waves.

Then, after extensive comparisons with numerical models, the team could determine a precise mass (0.586 solar masses), a rotation period of 1.38 days, and an upper limit of 6000 gauss on the surface magnetic field. In addition, they could conclude that the rotation and pulsation axes are aligned at an angle of 60 degrees to the line of sight, and that several thin layers of different chemical composition lie under the surface.

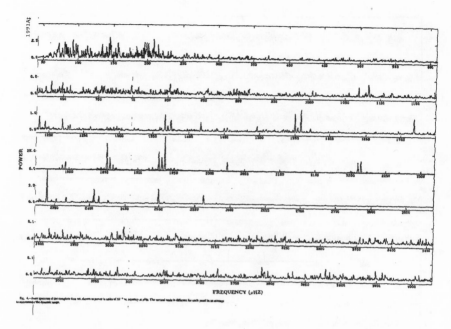

bar

FIG. 11.6   PG 1159-035 pulsates with over one hundred periods simultaneously. Each period is a peak in the frequency spectrum.

The team could justifiably feel proud of their accomplishment, but several pieces of the puzzle didn't fit into their neat picture. For example, they concluded that the star's pulsation periods were stable for years, but independent observations suggested otherwise. Four new periods of around 500 seconds suddenly appeared in 1987, in both the optical and X-ray light curves, and these were absent in the WET data. That raised the question of just how long a mode persists and how new modes appear. Furthermore, a basic question remained, namely, why only radial mode numbers (N) above 25 are excited. The team was able to circumvent this issue in their analysis, but it hovers out there still.

DEEPER AND DEEPER

As the WET team gained experience, they were able to extract more and more information about the interior structure of pulsating white dwarfs. For example, in 1990 the team observed the helium-rich DB white dwarf GD 358 and obtained 154 hours of nearly continuous high-speed photometry. They found no fewer than 180 periods in the pulsation spectrum in the range from 420 to 1000 seconds.

The periods were identified as $L = 1$ gravity modes, with high values of N. From the spacing of the peaks, a stellar mass of 0.61 solar masses was obtained, and from the variation of spacing, the mass of the outer helium shell was fixed at one-millionth of the stellar mass. The star's luminosity turned out to be only one-twentieth of the Sun's.

The most interesting discovery, though, concerned the rotation of the star. The team found strong evidence that the outer helium envelope rotates 1.8 times as fast as the core, that is, *differential rotation* in the radial direction. Something similar but less extreme, occurs in the Sun.

Finally, the team was able to detect a "weak" magnetic field of 1300 gauss. In comparison, the Sun's average magnetic field is less than one gauss.

LATE-BREAKING NEWS

The WET consortium has scheduled a campaign of one or two weeks, once or twice a year, ever since 1990. A large number of white dwarfs and similar objects have been examined, and the scientific results continue to pour in. In one of the latest campaigns, the team studied a pre–white dwarf, similar to the prototype PG 1159-035. This star was discovered by the European X-ray observatory ROSAT and turned out to have the hottest surface temperature (170,000 K) of any in its class. It was immediately checked for pulsations and found to be beating hard.

Fourteen telescopes joined as the WET during 1992, 1993, and 1994, accumulating as much as 175 hours each year of nonredundant data. This sounds impressive but still represents only about 6% of the possible total. Such a low "duty cycle" necessarily introduces spurious frequencies in the pulsation spectrum. Nevertheless, the team, led by French astronomer Gerard Vauclair, reaped a rich scientific harvest. They were able to identify thirty-seven of forty-eight gravity modes and from their frequency splitting determined a rotation period of 1.16 days. The amount of splitting decreased with the length of the period, which indicated that the surface is not rotating as a solid body; that is, they detected signs of *differential rotation in latitude,* similar to that of the Sun.

In addition, they were able to determine the star's mass (0.56 solar masses) and the fraction of that mass (about 0.05) in its helium shell. Finally, its absolute luminosity (four times that of the Sun) and therefore its distance from Earth (about 2500 light years) could both be fixed.

The white dwarfs have been a wonderful success for the young science of asteroseismology. As with the Sun, the combination of theory, modeling, and intensive observation has paid off in a detailed understanding of the interior of these tiny stars.

ANOTHER BREED OF CAT

White dwarfs are by no means the only stars that are known to pulsate; they are only the most intensively studied. Close behind in scientific interest are the Delta Scuti stars, indicated in figure 11.2 by the small circle at the bottom end of the instability strip. (Scutum is the Latin word for shield.)

These stars form a class of young variables, brighter, hotter, and slightly more massive than the Sun. They vary in brightness with periods between half an hour to eight hours. But unlike the white dwarfs, their brightness oscillates by only a few percentages, so they are more difficult to find. Nevertheless, they are surprisingly common. You find more and more of them as you search for weaker and weaker variables, and over 250 Delta Scuti stars have been found. In fact, Michel Breger, a pioneer in the field at the University of Wien (Vienna), estimates that 30% of all stars with their mass and surface temperature are variables.

Delta Scuti, which gave its name to this class of variable stars, was not the first of its class to be identified as a pulsator. That honor goes to L Monocerotis, an obscure little citizen of the constellation Monoceros, the unicorn. In 1974, two Australian astronomers discovered that this star varies in brightness with as many as eleven overlapping periods. At that time, the cause of the oscillations was unknown, but later work established that two of the strongest oscillations are nonradial pressure (acoustic) pulsations with $L = 1$ and $M = +1$ and -1, while the strong central peak is a purely radial pulsation.

A general rule seemed to hold for these stars: namely, those with large amplitude brightness variations pulsate in a few simple radial modes, while those with small amplitudes pulsate in several nonradial modes. The ones with radial modes have a nice feature: they obey a period-luminosity law, similar to the Cepheids, which makes them mildly useful in corroborating distances in the galaxy. The stars with several nonradial modes present a real puzzle, however, because theoretical models predict many more modes than are observed. Where are they?

In an attempt to find the missing modes and to obtain more precise periods, Breger and his associates organized the first network of telescopes in the late 1980s. This Delta Scuti Network, expanded and extended in longitude, has collaborated ever since, using high-speed photometers to follow the brightness variations of particular stars. The collaborators have had a number of notable successes in detecting multiple frequencies.

In 1995, for example, this Vienna-led network joined with the Whole Earth Telescope of white dwarf fame to study the Delta Scuti star FG Virginis. Nine observatories obtained a total of 170 hours of data. The light curve consists of ten periods between 42 and 150 minutes. The team used a stellar model with a mass of 1.8 suns to predict and identify the modes of these ten periods. They turned out to correspond to low-order acoustic and gravity modes, with $L = 0, 1, 2$ and $N = 1$ to 6.

In a follow-up to this campaign, six observatories observed the same star for a total of 435 hours over a period of forty days. This time the starlight in two colors was recorded, yielding nineteen periods, for a grand total of twenty-four. By comparing the difference in phase between the two colors with those calculated from a model, the team could make further L-mode identifications. The most important mode was the fundamental $L = 0$ radial mode, with a period of 119 minutes. Using this period as a constraint, many new models of the star were calculated and predictions of periods were compared with the data. In the end, Breger and coworkers were able to identify all twenty-four modes but could not find a model that would predict all periods exactly. The probable reason was the effects of rotation on the model.

With many more frequencies to work with, you might think that astronomers would have less difficulty in extracting physical information about these stars.

But there are at least two obstacles. The first and worst is that Delta Scuti stars rotate rapidly, with equatorial speeds as high as 200 km/s. Compared to the Sun's measly 2 km/s, this is a huge rotation and it complicates analysis, because rotation splits the observed frequencies into pairs and multiplets. In addition, fast rotation modifies the intrinsic *structure* of the star, so that a model must take it into account. That raises a catch-22 situation in which one needs to know the rotation to find a model and a model to identify the modes that reveal the rotation.

Several schemes have been proposed to get out of this box, and one of them may actually work. Most observations of Delta Scuti stars are measurements of varying brightness, usually in several colors. This method uses most of the available photons and therefore can get by with small telescopes. In contrast, spectroscopy of the stars uses only a tiny portion of the available light and therefore requires large telescopes, which are harder to mobilize in a network. But the pulsation of the star affects the width of spectral lines, and so spectroscopy offers another route to decoding the pulsation modes.

Michael Viskum and his associates at Aarhus University in Denmark have used this spectral line variation successfully on FG Virginis. From the ratio of the amplitudes of the line widths and the brightness of this star, the team was able to identify the eight strongest pulsation periods. Among these were two radial modes ($L = 0$) which are not affected by rotation, which allowed them to find a suitable model for the star. They could then predict its absolute luminosity and, by comparing that with its observed brightness, find its distance. Their estimate agrees very nicely with the distance (274 light-years) established by the Hipparchos satellite (see note 11.5). Finally, the optimum model yields the mass and mean density of the star, as well as some details on its internal structure.

One very puzzling and controversial feature of Delta Scuti stars is the variation in the strength (amplitude) of their modes. Michael Viskum and coworkers suggested that in FG Virginis, the variation can be explained by a rotation of 3.5 days. But in other stars, modes seem to come and go erratically. This could be a kind of "beating" among different periods or indicate a real change in the internal structure of the star.

Meanwhile, Michel Breger and his colleagues continue to rack up record numbers of pulsation frequencies. Their Delta Scuti Network has detected twenty-nine frequencies in BI Canis Minoris (the constellation of the smaller dog), twenty-two in XX Pyxidus, and a high-water mark of thirty in the star 4 Canem Venaticorum, the hunting dogs. But these are only a tiny fraction of the more than one thousand frequencies with low L that models predict. Where are all the others? The team was forced to conclude that some unknown selection mechanism suppresses the great majority of modes.

Two mechanisms have been suggested. Woitek Dziembowski, a well-known theorist at the Copernicus Astronomical Center in Warsaw, proposed "mode-trapping." In this scheme, an acoustic mode whose vertical wavelength just happens to match the thickness of a layer below the surface cannot escape from the layer and is almost undetectable at the surface. But although this scheme seems to work in white dwarfs, it only works in Delta Scuti stars for modes with $L = 1$. That's not good enough. Another possible mechanism goes by the daunting name of "parametric resonance." Suffice it to say that neither mechanism has been confirmed or rejected permanently. The puzzle of the missing modes remains unsolved.

Delta Scuti stars are intriguing and have spawned a small industry of observers and theorists. Several groups have taken different approaches toward securing good data. A Danish group at Aarhus University, for example, has used a two-site network (Small Telescope Array of CCD Cameras, or STACC) to look at several Delta Scuti stars in the same stellar cluster. Not only does this technique attain better use of observing time, but all the stars have similar composition and age, which simplifies the modeling. Moreover, the distance to the cluster is more easily determined than the distance to an isolated star (note 11.6). In 1998 this group observed two stars in the open cluster Praesepe with photometers and spectrographs and detected six modes in one star and four in the other.

A French-led network of telescopes is called Multi-Site Continuous Spectroscopy, or MUSICOS (bravo! this takes the acronym prize!). This group has also made important studies of Delta Scuti stars. As the name suggests, the group measures spectral line profiles and then Doppler shifts. In a typical study of 1992, a network of telescopes in China, Arizona, Hawaii, France, and the Canary Islands observed the star Theta Tauri for four days and nights. They discovered two pulsation periods that hadn't been seen before, periods that indicate that the star is unstable in more ways than one.

As with the Sun, the ultimate observing tool may be a satellite equipped with photometers and perhaps a spectrograph. In space the stars don't twinkle, a great advantage for detecting weak pulsations. In addition, a satellite is unaffected by weather and can observe its target as much as 90% of the time. That would yield pulsation spectra completely free of the spurious peaks that the day-night cycle introduces, and would lead to better detection of weak pulsations.

Plans for a dedicated satellite are afoot in several countries. The French are building a program called COnvection and ROTation of Stellar Interiors, or COROT. The Canadians want to launch Microvariability Of STars, or MOST. The telescopes aboard these vehicles are modest. But that may be good enough to break new ground. The European Space Agency is also planning a follow-on FLEX mission named Eddington.

With all this brainpower at work we can hope for the resolution of some of the outstanding questions in due course.

THE DEVIL WE KNOW: SUNLIKE STARS

Our Sun is a very common type of star, so there are plenty of candidates to examine for nonradial oscillations. Moreover, sunlike stars rotate slowly, so the problems of interpretation that arise with Delta Scuti stars are absent. However, as with our Sun, the oscillations are very weak and as a result the evidence obtained so far is ambiguous at best.

Perhaps the best example is the star Eta Boötis (Boötes is Latin for herdsman). It is somewhat more massive than our Sun, but has about the same surface temperature. In 1995, Hans Kjeldsen and his colleagues at Aarhus University observed the star for six nights with the 2.5-meter Nordic Optical Telescope. This telescope has a magnificent view from the rim of an extinct volcano on the island of La Palma in the Canaries. They used a powerful spectrograph to detect periodic variations of the strength of several hydrogen spectral lines. From the analysis of their data, they extracted thirteen modes with periods around twenty minutes. By comparing these periods with those predicted by Christensen-Dalsgaard, they concluded that the modes were acoustic modes, just as in the Sun. A nice result, indeed. But wait!

Tim Brown, the pioneer in helioseismology whom we have met before, has

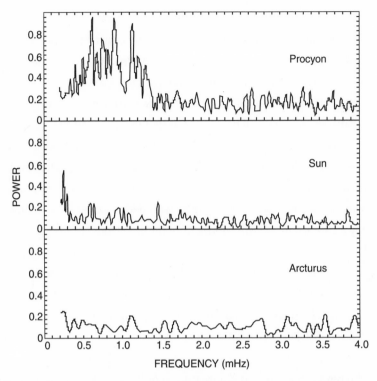

FIG. 11.7   Tim Brown and coworkers obtained this spectrum of oscillation periods for the sunlike star Procyon. For comparison, oscillation spectra of weak sunlight and of the star Arcturus are shown to demonstrate that Procyon does indeed pulsate.

moved on to search for oscillations in sunlike stars. In 1995 he and his colleagues attempted to follow up on the discovery of oscillations in Eta Boötis by searching for Doppler velocity variations. Doppler observations have several advantages over brightness measurements and so are worth pursuing.

The prospects for obtaining Doppler observations were not outstanding. Kjeldsen had estimated velocity amplitudes for Eta Boötis that were small (about 2 m/s) but, with care, detectable. So Brown and company used a sophisticated spectrograph to sum up the velocity signal from hundreds of spectral lines. They obtained twenty-two hours of data with this instrument. To their great surprise, they found *no evidence* for any of the thirteen periods that Kjeldsen had reported! Some hints of periods not mentioned by Kjeldsen were present, however.

Here was something of a mystery. Brown and colleagues puzzled long and hard about it and offered two possible explanations: either Kjeldsen made a mistake in estimating Doppler velocity amplitudes, or the data of one or both groups was contaminated by noise. The issue has never been resolved.

Procyon, or Alpha Canis Minoris, is a yellow star easily seen with the naked eye. It is somewhat hotter and heavier than our Sun but similar enough to expect solar-type oscillations. In 1988 and 1989, Tim Brown, Ron Gilliland, Bob Noyes, and Larry Ramsay had a look at it with a spectrograph. They searched for velocity oscillations of the star at the Kitt Peak National Observatory in Tucson. From ten nights of data they detected variations with periods around twenty minutes that they attributed to acoustic modes, as in the Sun. However, they were unable to isolate distinct modes (fig. 11.7). Brown and friends tried again in 1997 with an advanced version of their spectrograph, but with no better outcome.

In 1999 a French group headed by M. Martic obtained similar results for Procyon, with similar methods. Their theoretical colleagues, led by C. Barban, calculated a set of stellar models of Procyon and from these simulated the kind of time series that Martic and colleagues had obtained. Happily, the predicted series did resemble the observations, but it was quite clear that the daytime interruption of the series was corrupting the data. A multitelescope network would be needed to eliminate the day-night cycle from the data.

At last, in June 2001, two Swiss astronomers beat the day-night problem. François Bouchy and Fabien Carrier detected unambiguous acoustic oscillations, with periods near seven minutes, in the solar-type star Alpha Centauri A. They used an extremely stable spectrometer during five nights at the Swiss telescope at the European Southern Observatory in Chile. Then in May 2002, an international team detected oscillations around three hours in Xi Hydrae, a red giant star unlike the sun. Asteroseismology is well on its way!

SOME LATE NEWS 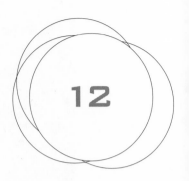 12

WHEN YOU MOVE into a new neighborhood, you are most concerned with finding your way around. You look for signposts, landmarks, and patterns of any kind. You gradually get a map of the territory in your head. If you live there long enough you begin to notice the smaller details, a broken window, a flag. Finally, you notice changes, a tree coming into leaf or a new display in a storefront.

In much the same way, solar helioseismologists have gradually grown familiar with the interior of the Sun. At first they were able to map only the grossest features of the interior. Eventually, with long, uninterrupted time series of observations in hand and with refined techniques, they have been able to resolve finer features as well as slow changes inside the Sun. In this chapter we will learn what they have been up to during the past few years.

A REAL SUNQUAKE

As everyone knows, seismologists probe the interior of the Earth by studying the pressure and shear waves that earthquakes produce. Helioseismologists have learned the same trick, except that the waves they study in the Sun aren't produced by some catastrophic event, but rather by the steady noisy process of convection.

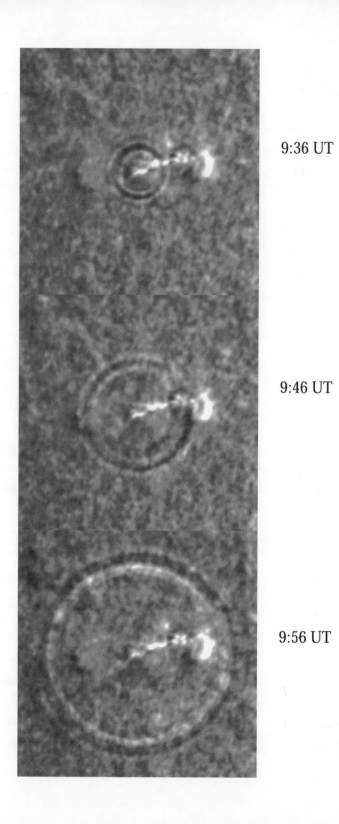

9:36 UT

9:46 UT

9:56 UT

Nevertheless, astronomers wondered whether some dramatic event, perhaps a solar flare, might kick off a train of waves that would penetrate into the Sun. A solar flare, as you know, is a massive explosion in the low corona. A huge amount of energy, sometimes equivalent to a million megatons of TNT, can be released in a few minutes. The plasma is heated, sometimes to tens of millions of degrees, and expands violently. If a shock wave from the flare struck the Sun's surface, it might send seismic waves propagating into the interior.

Nobody had seen such an event, however. Deborah Haber and Frank Hill, two scientists at the National Solar Observatory in Tucson, Arizona, looked for one in 1988. They had seen a powerful flare occur in the middle of their oscillation observations, so they searched for evidence of traveling waves before and after the flare. Nothing definite was seen.

Then in 1998, Alexander Kosovichev of Stanford University and Valentina Zharkova of Glasgow University succeeded. A powerful flare, with enough energy to power the United States for twenty years, occurred on July 9, 1996. The MDI aboard the SOHO satellite captured the event, but not until the two researchers hunted through the data, in 1998, did it come to light. A beautiful series of circular surface waves spread out from the flare (fig. 12.1). The event resembles the ripples set off by dropping a pebble in a pond. But unlike the pond ripples, which travel at constant speed, these waves accelerated from 10 km/s to over 100 km/s before disappearing.

Kosovichev and Zharkova proposed an explanation for the event. A flare fireball would have produced a pattern of waves far more complicated than was observed. So they proposed instead that the cause of the wave was the impact upon the surface of the Sun of a beam of high-energy flare electrons. Such focused beams are well known to flare researchers. They produce bursts of X rays and cause the chromosphere to heat up impulsively when they strike the surface. But this is the first documented case of a flare causing a sunquake.

Facing page:
FIG. 12.1   On July 9, 1996, a powerful solar flare set off these circular waves over the solar surface.

THE NOT-SO-CONSTANT SOLAR CONSTANT

When somebody invents a new tool, he or she can occasionally make an important but unexpected discovery with it. A good example is Hans Lippershey, the Dutch maker of spectacles, who in 1608 put two lenses in a tube and invented the telescope.

Richard Willson had a similar experience. He is a talented physicist who was employed at the Jet Propulsion Laboratory in Pasadena, California. For the past thirty years he has been working to perfect instruments for "remote sensing" of processes in the Sun and the Earth's atmosphere. In 1980, he developed an extremely sensitive device to measure the total amount of the Sun's radiation, the "solar constant," which is an essential factor in modeling the Earth's climate. Willson's brainchild, the Active Cavity Radiometer Irradiance Monitor (ACRIM), was capable of detecting changes in the amount of sunlight reaching the Earth as small as one part in a million. The first version of the ACRIM was installed in the Solar Maximum Mission, a solar satellite that orbited the Earth for a decade.

The ACRIM soon proved its worth. It revealed that when a large dark sunspot appears on the face of the Sun, the amount of sunlight received by the Earth can drop by as much as 0.2% in a week. This news sent the theorists scrambling, looking for an explanation (note 12.1).

As the years went by, the signal from the ACRIM began to droop, and Willson was hard-pressed to explain the change. Fortunately, he had provided three copies of the ACRIM on board the SMM satellite and could cross-check among them. He concluded that the weakening signal was not the fault of his instruments but rather a real change in the Sun. He patiently collected the data, and after seven years he and his colleague, solar physicist Hugh Hudson, concluded that the total radiation of the Sun varies in step with its eleven-year magnetic cycle.

Figure 12.2 shows this result. The intensity of sunlight reaching Earth varies erratically as sunspots come and go, but the long-term trend is unmistakable. The Sun is brighter at sunspot maximum by about 0.1% than at sunspot minimum. This is indeed an astounding result, one that astronomers are still trying to explain fully.

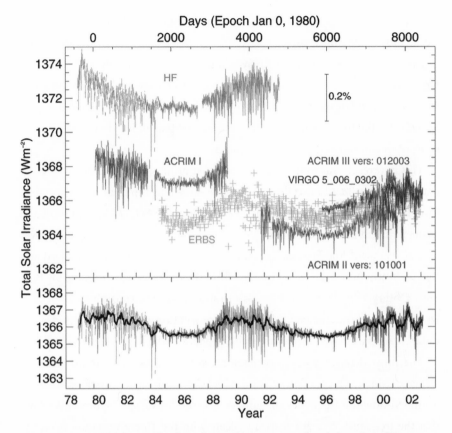

FIG. 12.2   ACRIM I was the first of several instruments placed aboard satellites to measure the Sun's total output of radiation, the "solar constant." As dark sunspots and bright active regions rotate across the solar disk, the "constant" fluctuates. In addition, the Sun is actually about 0.1% brighter at sunspot maximum than at minimum.

EVERYTHING CHANGES, IF YOU LOOK LONG ENOUGH

There was more to come. In 1985 Martin Woodard, a graduate student at the University of California at San Diego, and Hugh Hudson, his thesis adviser, discovered that the five-minute oscillations of the Sun show up as periodic fluctuations in the brightness of sunlight. From ten months of ACRIM data they could detect the low L-modes, L = 0, 1, and 2. The amplitude of the oscillations was weak, only a few parts per million, but the ACRIM was sensitive and stable enough to make a firm identification possible. Moreover, Woodard and Hudson were able to conclude that these low-degree modes maintain constant phase for at least a week before changing abruptly.

Martin Woodard and Robert Noyes, a professor of astrophysics at Harvard University, took the next step in the saga. In 1985, they analyzed five years of ACRIM data and discovered that the *frequencies* of the low L-modes had changed. Specifically, the frequencies had decreased by one part in ten thousand between sunspot maximum and sunspot minimum.

This new result was confirmed later by other groups, who used quite different data. Tom Duvall and his colleagues saw the same effect in their oscillation data from the South Pole. All the intermediate L-modes, up to L = 150, had shifted in frequency. Tim Brown saw the frequency changes in his velocity data, taken over several years with the Fourier tachometer (see note 5.3). In 1993, Woodard and Ken Libbrecht found the same effect in Libbrecht's velocity data from the Big Bear Solar Observatory. They were even able to show that the changes were associated with a one-percent change in the subsurface rotation rate at high latitudes. All these changes were well correlated with the phase of the solar magnetic cycle.

The most detailed study was made recently by Thierry Corbard and colleagues. Using six years of Steve Tomczyk's LOWL velocity oscillations, they determined how the frequency of individual L-modes varied. Figure 12.3 shows that the frequency changes correlate nicely with the 10 cm radiation from the Sun, a common index of solar activity. Here indeed was a clean-cut problem for the theorists to tackle.

The theorists rose to the occasion. Peter Goldreich and his colleagues at Caltech made the first attempt in 1991. They investigated whether changes in the temperature or magnetic field strength in the interior could reproduce the observations. A change in the temperature profile in the interior tends to lengthen the path lengths of sound waves and to *reduce* their frequencies. On the other hand, an increase in the mean magnetic field in the interior tends to raise the propagation speed of sound waves and therefore to *increase* their frequencies. The shift to higher frequencies observed during the rise of the solar cycle pointed directly to the magnetic field as the cause. A mean field strength of only 200 gauss was sufficient, they estimated, to account for the observed rise.

Then in 2001, Douglas Gough, that irrepressible theorist at Cambridge Uni-

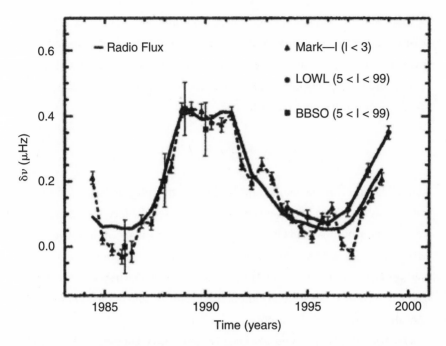

FIG. 12.3 Oscillation frequencies shift slightly in step with the solar cycle. This figure compares the average shift, from different experiments, with the Sun's 10 cm radiation, which is a good measure of solar activity.

versity, reexamined the problem, along with a stellar cast of colleagues. In the ten years since Goldreich's work, the Birmingham group (BISON) had discovered that the *widths* of the peaks in the oscillation spectrum had increased by 24% from 1991 to 1997. These widths are direct measures of the lifetimes of different modes (see note 12.2).

Gough and his colleague George Houdek investigated how the growth of magnetic fields in the Sun's interior could affect the horizontal sizes of convective flows, and how these in turn could affect both the oscillation frequencies and oscillation line widths. They made numerical simulations of convection, with a modern treatment of turbulence. When they compared the calculated and observed frequency changes, they found a reasonable fit. So it seems that the growth of magnetic fields actually changes the structure of the Sun, and those changes affect the oscillation spectrum.

Changes in the oscillation spectrum offer one more way to investigate how the solar cycle works. The long-term databases now available are enabling astronomers to track subtle variations in convection patterns and the interaction of convection and magnetic fields. This is an active field now, with conferences and research papers burgeoning everywhere.

RIVERS ON THE SUN

Robert Howard is an old hand in the business of mapping the magnetic and velocity fields of the sun. For over twenty years, he used the forty-five-meter solar tower at the Mount Wilson Observatory to make daily maps of these two essential quantities over the full disk of the Sun. Howard is a very careful observer and has the patience of a stalking cat. His patience has paid off. His precious database has enabled him to extract many important characteristics of the rotation of the Sun, and in particular, of sunspots.

Way back in 1980, he and graduate student Barry LaBonte had examined twelve years of data for changes in the average rotation speeds at different latitudes at the surface of the Sun. They found several long-term variations as well as a curious tendency for the variations to be different in the Northern and Southern Hemispheres. But in addition, they were left with a puzzling residual in their analysis. The Sun rotates at about 2 km per second at the equator, and these residuals were tiny by comparison, a mere two or three *meters per second.* Anyone else would have discarded this remnant as an annoying bit of noise in the data.

But Howard persisted. He knew precisely how accurate his data were, and they were reliable, he felt, to better than a meter per second. So he and LaBonte looked more carefully. They discovered four east-west bands in each hemisphere in which the rotation rate alternated between 2 m/s faster and 2 m/s slower than the average for that latitude (fig. 12.4). Even more interesting, these bands first appeared at high latitudes in both hemispheres and migrated toward the equator in a period of about twenty-two years, or twice the classical period of the sunspot cycle.

Howard and LaBonte concluded that they had found a large-scale, deep-

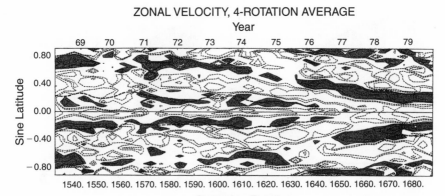

ZONAL VELOCITY, 4-ROTATION AVERAGE

FIG. 12.4 Alternating bands of slightly faster and slower rotation drift toward the equator in about twenty years. One Carrington rotation equals twenty-seven days.

seated oscillation of the Sun. Because it was symmetrical in both hemispheres, it had to be a global phenomenon, not a local fluctuation. And the characteristic period suggested that these bands must somehow be related to the solar cycle.

Eighteen years later, Jesper Schou at Stanford University and a group of twenty-three coworkers presented a detailed analysis of 144 days of continuous MDI observations. Their principle result was a superb map of solar rotation as a function of depth and latitude (fig. 12.5). In addition, they were able to probe the migrating bands discovered at the surface by Howard and LaBonte. They determined that these bands (fig. 12.6) have speeds about 5 m/s faster or slower than average and may persist as deep as 30,000 km below the surface.

Everybody agrees that these bands are probably important clues to the operation of the solar dynamo, but as yet no one has provided a satisfying explanation for them. As noted earlier, theorists are still struggling to explain the basic map of rotation as a function of depth and latitude. Stay tuned.

YET ANOTHER PERIOD IN THE CYCLE

Rachel Howe has had a short but varied scientific career. She started out as a physics major at the University of Birmingham in England, and as a graduate stu-

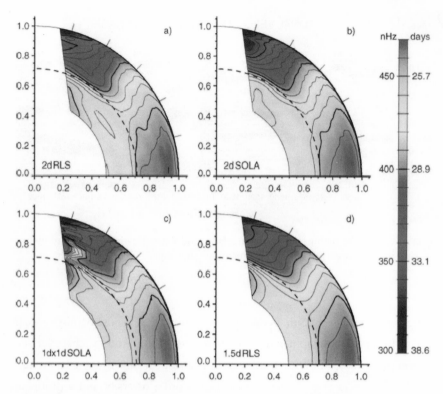

FIG. 12.5 The interior rotation of the Sun, determined by MDI data. Four different methods of analysis yield very similar results and set limits on the precision of the map.

dent reduced data for the BISON oscillation group. Her first love was solid state physics, however, and in 1988 she completed her doctoral dissertation, "Ice IX, the Ordered Form of Ice I h." Then she was apparently lured back into helioseismology. For seven years she worked as a research assistant in the BISON group, joining in as a coauthor on several papers, but never making her own major discovery. In 1997 she migrated to the GONG project at the National Solar Observatory in Tucson and in the year 2000, she hit a home run.

She analyzed three years of GONG and MDI data in order to determine the rotation speeds in the vicinity of the tachocline, the thin layer at a depth of 0.3 radius that separates the convection and radiative zones. As we have learned, the

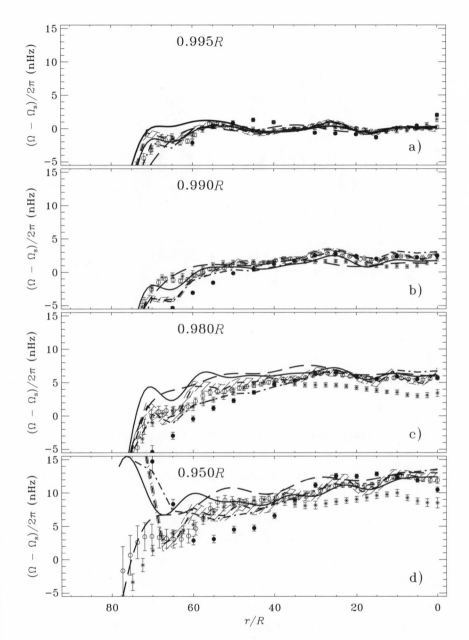

FIG. 12.6   Alternating bands of slower and faster rotation extend at least 35,000 km into the Sun. These bands appear clearly in observations from the MDI aboard SOHO. The panels show the differences of rotation as functions of latitude and depth.

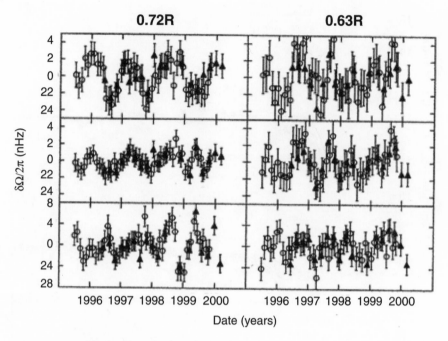

FIG. 12.7 The rotation speeds at the top (0.72R) and bottom (0.63R) of the tacho-cline oscillate with a fifteen-month period. The largest amplitude occurs at the equator.

rotation speed varies from pole to equator along the upper surface of the tacho-cline and is nearly constant from pole to equator along the lower surface. At the equator of the tachocline, there is a difference of about two days in the period of rotation between the top and bottom surfaces. Howe easily confirmed this pre-vious result with the long series of observations available to her.

But a completely unexpected effect dropped out of her analysis: the difference of rotation periods across the thickness of the tachocline *oscillates by about 20% in a period of fifteen months.* Both the independent data sets from the GONG and the MDI revealed this curious oscillation.

Thirty months later, she had data from both instruments for two additional oscillations. Figure 12.7 shows how the two data sets agree in defining the oscil-lation of rotation periods, at both the top (0.73 radius) and bottom (0.63 radius) of the tachocline. The changes are largest at the equator but are certainly present at latitude 60 degrees.

This result is too new and too startling to have produced an explanation yet. But have no doubt that the bright minds out there are working on it!

LISTEN TO THE DYNAMO HUM

I've saved the best for last. As of 1999, great progress is being made in understanding how a solar dynamo, deep inside the Sun, might produce the solar cycle effects we see at the surface. We are still far from a complete dynamical theory that would explain the Sun's internal rotation as well as its magnetic cycle, but if one adopts the observed map of rotation, one can make some headway with the dynamo. And this is what several groups around the world have done. We focus here on the work of the group at the High Altitude Observatory in Boulder, Colorado.

We have met two members of this group already, Peter Gilman and Paul Charbonneau. Gilman is an elder statesman among helioseismologists, who has been hammering at the dynamo problem since the 1980s. Charbonneau is a young theorist with a good track record. When they were joined recently by Mausumi Dikpati, a theorist from the India Institute of Astrophysics, they began to make spectacular progress.

First, remember that the goal of dynamo theorists is a physical mechanism that would convert "poloidal" magnetic fields (like those around a bar magnet) into "toroidal" fields (shaped like doughnuts) that wrap around the Sun, *and vice versa,* in an eleven-year cycle.

Let's recall the status of the subject as of 1998. A new map of the internal rotation of the sun (see fig. 12.5) showed that the rotation of the convection zone varies in latitude but not along a radius. That meant there were no shearing motions, which are essential for converting poloidal to toroidal fields, in the convection zone. And that meant that the dynamo had to reside elsewhere. The most likely place was in the tachocline, the thin transition layer between the differentially rotating convection zone and the rigidly rotating radiative zone (see fig. 10.8). At the tachocline, the two zones rub against each other and provide those essential shearing motions.

The tachocline has other advantages. Theoretical calculations had shown earlier that only magnetic ropes with very strong fields (60,000 to 160,000 gauss) are able to float to the surface at the observed latitudes and with the observed tilts of pairs of sunspots (see chapter 10, p. 185). To allow time for such strong fields to grow, the ropes have to remain in the dynamo region for years. That is indeed possible just below the tachocline, in the radiative zone, where the ropes are not buoyant.

How would a dynamo work in the tachocline? First, Gene Parker showed how a dynamo could behave at a very sharp "interface" at the base of the convection zone, under certain conditions. Later, Paul Charbonneau and Keith MacGregor demonstrated one example of a dynamo model based on this idea (see chapter 10). But the details of the dynamo within a realistic tachocline remained to be worked out.

A number of theorists, including Gilman and Charbonneau, began to worry about the *stability* of such a thin tachocline. Why wouldn't the shearing motions at the top of the tachocline propagate down into the radiative zone and cause it to rotate differentially? At first the theorists couldn't agree whether the tachocline was stable, with or without a magnetic field. The final conclusion was that it was indeed stable but rippled with a type of hydromagnetic wave.

At this point (around November 2000), Peter Gilman had a brilliant idea. He saw an analogy between waves on the surface of a pond and the kind of waves that seemed to ride on the tachocline. He modified the water wave theory to take into account the presence of strong magnetic fields. Then he and Dikpati applied the modified theory to the tachocline. In their model, they allowed the tachocline to oscillate radially as well as longitudinally, in a kind of weak three-dimensional pulsation. And they made a wonderful discovery.

The pulsations of the tachocline can produce *vortical* motions, like the whirl-pools that an oar leaves behind as a person rows across a pond (fig. 12.8). These vortical motions would be ideal for twisting a magnetic rope as it slowly diffuses through the tachocline toward the surface. The twist would then show up as the tilt in the axis of a pair of sunspots. Or, in other words, the tachocline could be the source of the so-called alpha effect that formerly had been attributed to cyclonic convection cells (see chapter 10, p. 184).

So the tachocline was now equipped to carry out both functions of a dynamo.

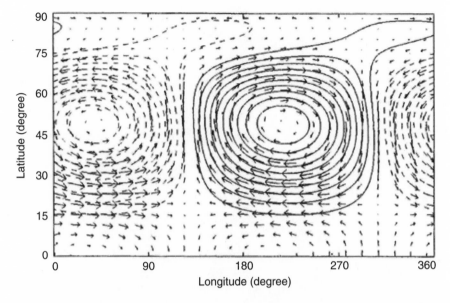

FIG. 1 2.8   Waves on the tachocline produce large-scale vortices and the alpha effect, according to a model by Mausumi Dikpati and Peter Gilman.

It could *amplify* toroidal fields by wrapping them around the Sun (the "omega effect"), and it could also *twist* these magnetic ropes by the proper amounts (the "alpha effect") to regenerate the poloidal fields.

One cog in the dynamo machine was missing, however. The poloidal fields at the surface have to be carried down to the tachocline, where they can be converted to toroidal fields. How could this happen? Bernard Durney at the University of Arizona and others had postulated that a slow conveyor belt in the convection zone does this job. At the surface, the plasma is observed to flow toward the poles at a mere 10 or 20 m/s, carrying the magnetic fields with it. (This is called "meridional flow.") Then to complete the circuit, the flow has to dive down at the poles, stream equatorward along the top of the tachocline, and rise again.

Now the task was to put all this together into a new model of the solar dynamo. Dikpati and Gilman offered such a model in September 2001. The model is kinematic, that is, it adopts the observed map of the Sun's internal rotation as a starting point, without trying to generate it from first principles. Figure 12.9A

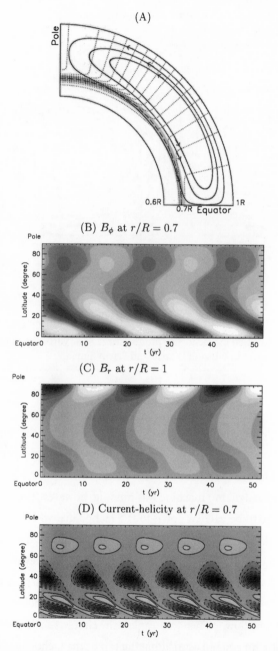

FIG. 12.9 The output of a tachocline dynamo in the north solar hemisphere. (A) The assumed pattern of meridional flow; the shaded area is the tachocline. (B) The toroidal magnetic field strength at the top of the tachocline. The lower half of each panel resembles the "butterfly diagram" of sunspots; the field varies with a ten-year period and the maximum drifts toward the equator. (C) The predicted radial field strength at the surface shows the observed migration to the equator and to the pole. Panel D shows the variation of the vortical motions on the tachocline.

shows the velocity patterns the model employs, and the other panels illustrate how the magnetic fields vary during a cycle. In panel B of this figure we see how the strength of toroidal fields vary in the Sun's north hemisphere at the top of the tachocline. With the surface meridional flow set at a realistic value of 17 m/s, the model predicts that the fields will migrate to the equator in about ten years. In short, the model can reproduce the famous "butterfly diagram" of sunspots in the north hemisphere. In the next ten-year period, the field polarity reverses, just as Hale's rules require. Panel C of the figure shows the radial magnetic field at the surface, the quantity that observers can actually measure. Here again we see the migration to the equator. In addition, some fields migrate toward the pole, which is required to reverse the polar field polarity. A nice feature of the model is that it even predicts the correct phase of the cycle at which this reversal takes place.

This model worked so well that Dikpati and Gilman decided to extend it to a complete spherical shell that would include the north and south hemispheres simultaneously. When they did that, they made a very interesting discovery.

Recall that, according to Hale, the leader spots in the north and south hemispheres have the opposite polarity during one cycle (see fig. 10.4) and then flip polarities in the next cycle. In other words the polarities are *antisymmetric* about the equator.

As figure 12.10A shows, the Dikpati-Gilman model, with the alpha effect operating in the tachocline, predicts this antisymmetry nicely. The big surprise was that conventional Babcock-type models, with the alpha effect operating in the body of the convection zone, *do not.* Such models predict *symmetric* polarities in the north and south hemispheres, as figure 12.10B illustrates. Dikpati and Gilman argue that this difference in behavior makes a strong case for their tachocline dynamo.

How could so many modelers have missed the essential flaw in assuming an alpha effect in the convection zone? One strong possibility is that, to save on computation, polarities in the south hemisphere were simply defined to be mirror images of those in the north, a very reasonable assumption.

Gilman and Dikpati emphasized that the velocity of the meridional flow is a critical factor in determining the period of the cycle. Now, as we have learned previously, the average period of the cycle is about eleven years, but cycles as short

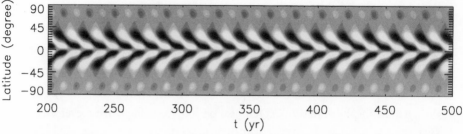

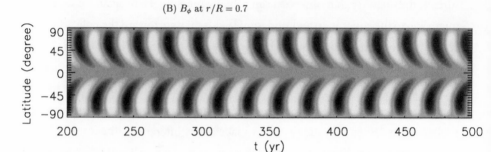

FIG. 1 2 . 1 0 (A) The Dikpati-Gilman model predicts correctly the antisymmetry of magnetic fields in the north and south hemispheres, while (B), the Babcock model, does not.

as eight years and as long as twelve are common. In addition short cycles produce higher peaks in the number of spots than long ones.

When Dikpati and Charbonneau introduced random variations of the meridional velocity in a model, they discovered to their delight that they could reproduce this well-known property of the solar cycle. Short and long cycles occur, with a tendency for a long one to follow a short one, but the long-term average is fixed at eleven years by the long-term average velocity of about 17 m/s.

All this looks very promising. The next question might be, how could one check up on this tachocline model of the dynamo? Dikpati and Gilman provided one possible test. They predicted that the tachocline should have a "prolate" shape like a football, with a longer diameter at the poles and a smaller diameter at the equator.

In 2001, Thierry Corbard and his coworkers at Cambridge University analyzed six years of data from Steve Tomczyk's LOWL instrument to investigate the precise shape of the tachocline. These observations gave them a resolution of a tenth of a solar radius in depth (70,000 km) and about 20 degrees in latitude.

In each year of data they found indications that the tachocline does have a prolate shape. However, the measured difference between the polar and equatorial diameters was about as small as the uncertainty in the thickness of the tachocline itself, .03 of a radius, and therefore not significant. No doubt this test will be repeated to get more complete data.

Kinematic dynamos, which adopt the observed pattern of solar rotation, have come a long way since the tachocline appeared on the scene. However, these models cannot predict the actual strength of the magnetic fields, only their variations. The next big step will be to combine the physics of convection and differential rotation (see chapter 9) with a tachocline dynamo, to create a truly satisfying understanding of the solar cycle.

EPILOGUE:
WHAT'S NEXT?

HELIOSEISMOLOGY has been enormously successful in revealing the internal structure of the nearest star. Although many details remain to be settled, the broad outlines of the average properties of the Sun have been determined. The temperature profile, the pattern of rotation, the chemical composition—all these have been nailed down more securely by means of the solar oscillations. Large-scale internal motions, such as torsional oscillations and meridional flow, have been mapped. Smaller objects, such as sunspots and active regions, have also been explored. And the generation of the Sun's magnetic fields during an eleven-year activity cycle has been clarified, if not completely resolved.

The new observations have generated a powerful surge in theoretical and numerical studies. Indeed, one of the most satisfying aspects of the science has been the close interaction of theory and observation, and the convergence of several subfields in solar physics. Such examples as the simulation of convection, rotation, flux rope dynamics, and the solar dynamo come to mind.

To a large extent, the initial goals of helioseismologists have been met. What's next? This question is being debated in the community of scientists. Some outlines of possible future work are becoming clearer. Some of the big problems, such as the detailed operation of the solar dynamo, remain to be solved, which will entail a much closer examination of the dynamics of the tachocline. Similarly, we still do not have a complete understanding of the origin of the Sun's complicated pattern of rotation, which will undoubtedly spark new investigations and simulations of convection. Asymmetric flows, associated with the emergence of active regions, for example, lack a physical explanation so far.

Many questions that are associated with variations of internal structure on timescales of a month to a decade remain to be answered. We still have a poor understanding of how the oscillations are generated and how they decay, for example.

Furthermore, there are regions that have been incompletely explored, such as the thin layer just under the surface, where a radial gradient of rotation speed exists. Does this layer generate magnetic fields, and if so, how? The high latitudes and the deep core are also still incompletely mapped. And the backside of the sun, which only recently has become visible in sound waves, can teach us more about the birth of active regions.

At least two major goals have eluded the observers so far: the detection of gravity modes, which probe the deep interior; and of strong subsurface magnetic fields, which ultimately emerge to create all the phenomena at the surface. There are good theoretical reasons to expect that detection of either the modes or the fields may be impossible, but one never knows.

The next generation of seismic instruments is being planned with all these questions in mind. At least one is well along. NASA is planning to launch the Solar Dynamics Observatory, a second-generation SOHO, in 2007. The satellite will focus on the Sun's influence on the Earth's climate and space environment. It will carry a complement of instruments to measure solar oscillations, magnetic field, solar wind, and flare emissions. The SDO will be designed to remain in geosynchronous orbit 22,000 miles above the Earth for at least five years, providing a constant stream of data about the complex magnetic fields generated deep in the solar interior.

On August 23, 2002, Phillip Scherrer of Stanford University, a pioneer in helioseismology and principal investigator for the MDI aboard SOHO, was chosen to lead a team of scientists to investigate velocity and magnetic fields below the Sun's surface. They will build the Helioseismic and Magnetic Imager, an oscillation detector that will generate 3-D images of the solar interior.

The European Space Agency, ESA, is in the early stages of planning the Solar Orbiter, a successor to the Ulysses satellite. The Orbiter will approach the Sun to within two-tenths of the Earth-Sun distance to make measurements of the solar wind and to examine the magnetic fields at the surface at high resolution. Whether an oscillation detector similar to the MDI would be included depends on a sharper definition of the science objectives.

Asteroseismology is still in its infancy. With the launching of several small satellites, dedicated to the subject, we can hope for more rapid progress in this field as well.

In the end, one cannot really predict where the next exciting developments will occur. Who could have predicted the fifteen-month variation of oscillation frequencies, for example? What will more precise oscillation frequencies reveal next month or over the remainder of this solar cycle? We shall just have to wait and see.

NOTES

1 • THE DISCOVERY

1.1 The Coronagraph

Bernard Lyot's coronagraph was a refractor with a single front lens. That lens had to be nearly perfect, with no scratches or dust or internal flaws that would scatter sunlight from the bright center of the solar disk to the ring just outside the disk where the faint corona might be seen. The lens formed an image of the solar disk in front of a metal "occulting disk," or artificial Moon, which blocked all the light of the disk but allowed the light in the ring outside the disk to pass. This coronal light was then refocused on a photographic plate.

1.2 The Doppler Effect

Imagine a water bug tapping some of its feet in unison once per second as it rests on top of a pond. The bug's tapping will send out circular wave fronts, one per second, separated by a fixed distance, the wavelength. Now imagine that the bug moves toward you across the water at a speed of a quarter of a wavelength per second while continuing to tap at a steady rate of one per second. The wave pattern you will receive is now crowded together by the bug's forward motion and will have a higher "pitch," with a wavelength one quarter shorter than before. The pitch of the bug's signal has been shifted to a higher tone because of the bug's movement toward you. This Doppler effect occurs in all wave phenomena, including light and sound. The rise in pitch of a police siren as a patrol car approaches is another example.

1.3 Averaging Velocities

A simple arithmetic average of velocities would always yield a result close to zero, because at any moment as many granules have negative velocities (are rising) as have positive velocities (are sinking). Evans therefore squared each granule's velocity, took the average of the squares, and then took the square root of that mean to get the root-mean-square value, as shown in figure 1.4.

1.4 Probing the Solar Atmosphere

Looking into the Sun's atmosphere is like looking into a fog. At most wavelengths, the fog is relatively transparent and one can see deep into the atmosphere. At other wavelengths (such as the centers of Fraunhofer lines), the fog is more opaque and one can see only down to a shallow depth because some types of atoms absorb light strongly at those wavelengths. The stronger (i.e., darker) the Fraunhofer line, the more opaque the atmosphere and the higher the level one sees. Therefore, by choosing Fraunhofer lines with different strengths, Evans could sample different heights in the Sun's atmosphere.

1.5 The Spectroheliograph

This instrument combines a spectrograph and a moving photographic plate (see fig. 1.5). An image of the Sun is formed on a linear slit (S1). The slit admits light to the machine from a narrow slice of the Sun that is focused on a diffraction grating (G). The grating spreads out sunlight into its constituent colors and Fraunhofer lines. A second slit (S2) is positioned at the desired wavelength within a single Fraunhofer line. The light that passes through this second slit is focused on a photographic plate that is mounted on a moving stage. As the solar disk is stepped across slit S1, the plate is moved in tandem, exposing a fresh part of the emulsion at each step. In this way, an image of the Sun is constructed, line by line, in the light of the selected Fraunhofer line.

1.6 Subtraction of Photographs

Robert Leighton improved a method first used by Fritz Zwicky, an extragalactic astronomer. To subtract one photograph from another, Leighton first prepared a positive contact transparency of one of them and developed it to a gamma of unity, reproducing the differences in light intensity by equal differences in transparency. He could judge whether he had achieved unity gamma by overlaying the transparency with its original negative and requiring the pair to look uniformly gray. Then he overlaid the transparency with the second original photograph and made a new photograph from that sandwich. All solar features (like sunspots) that are the same on both originals cancel in this final plate, and only the differences, which are displayed as bright or dark features against the gray background, remain.

1.7 Measuring Doppler Velocities

In figure 1.7, the solid line indicates the position in wavelength of a Fraunhofer line when its source is at rest with respect to the observer. If the source is receding, the line shifts to the right, toward longer wavelengths, as shown by the dotted line. The amount of the shift can be determined by measuring the difference in the amount of light at the two fixed wavelength bands in the wings, marked a and b. Note that this difference is zero for the source at rest and nonzero for the moving source. The shift is easily converted to velocity by a simple formula.

1.8 Delay Times

Recall how a spectroheliograph works. The solar image is stepped across a long straight slit, and a photograph of the slice of the sun that passes through the slit is taken at each step. The photographic plate is moved between steps to reveal an unexposed surface, and so the photograph is built up in monochromatic light, slit width by slit width.

In Robert Leighton's procedure, each point on the solar image was photographed twice: on the first plate when the entrance slit passed it in the forward direction, and on the second plate in the reverse direction. At one edge of the disk, where the forward scan stopped, the reverse scan was started immediately, that is, with *no* time delay. Points on the opposite edge, in contrast, were first photographed at the *beginning* of the first scan and then at the *end* of the second scan, with a time delay of twice the scan time, or about ten minutes. Between the edges, the delay between exposures increases linearly, from zero to a maximum of two scan times. In this way, corresponding points on two Dopplergrams are associated with a definite delay time.

2 • CONFUSION AND CLARIFICATION

2.1 A Primer on Waves

A wave is a periodic displacement in some kind of medium, like air or water or solar gas. If you lie on a raft on the ocean, you can feel the waves going past by the rise and fall of the ocean's surface. In such a *surface gravity* wave, the water oscillates perpendicular to the direction in which the wave is traveling. In a *sound* wave the oscillations of air are in the *same* direction in which the wave is traveling.

Strike a note on a wine glass and the blob of air near your ear rocks back and forth along the direction from wine glass to ear. But no blob of air travels from the glass to your ear like a little bullet; instead, each blob travels only a tiny distance, back and forth. Only the sound energy travels the whole distance, passed from one blob to the next, like a bucket in a bucket brigade, going at the speed of sound in air, about 350 m/s at room temperature. For comparison, the speed of sound in the Sun is about 10 km/s near the surface and increases inward as the square root of the temperature.

A sound wave is characterized by its wavelength (L, the distance between maximum displacements) and its frequency (F, the number of times a second the maximum displacement occurs at any chosen point in space). In a sound wave, the two quantities are related by the speed of sound S: $S = F \times L$. The wave's period P is the time it takes to complete a full cycle of the oscillation, and $P = 1/F$. If a wave travels at an angle to the horizontal, its wavelength can be split into a horizontal and a vertical component.

If we focus attention on a little blob in the medium at any fixed point in space, we see the blob oscillating about the fixed point. The maximum displacement of the blob from the chosen point is the wave's "amplitude." The "phase" of the oscillation describes where it is in its cycle. (Recall that we speak of the phases of the Moon.) At phase zero, the displacement of the blob is zero; it is passing through the fixed point with maximum speed.

At a phase of a quarter of a period later, the displacement is at its maximum and the speed of the blob is zero. And so on.

2.2 Traveling Waves, Standing Waves, and Resonant Cavities

Traveling waves. If you place dominoes on their ends in a row, and then tip over the first one, the second one falls, tipping over the third and so on. A kind of wave (a pulse) travels along the entire row at high speed, while each domino moves only a short distance. This is an example of a traveling wave.

Strike a tuning fork and you will hear a pure note, a traveling sound wave with nearly a single frequency. At the top of figure 2.6, we see a short segment of the wave at its beginning and a quarter cycle later. Note how the crests and valleys of the wave move as a whole; the entire waveform is simply displaced. Another way of expressing this is to say that the *phase* of the wave is traveling. A crest, for example, which is at a particular point in space, passes next to the neighboring point, and so on. The phase at any point varies periodically in time, and the phase at any instant varies periodically with distance along the wave.

Standing waves. Now imagine that the two waves shown at the top of figure 2.6 are traveling in opposite directions. Where they overlap, their amplitudes add. If these two waves are displaced in phase by half a wavelength, their sum will be a standing wave, whose nodes do not move. In a standing wave (fig. 2.6, bottom) the phase is the same everywhere at any instant, but it changes periodically. The amplitude, in contrast, varies sinusoidally from point to point.

Next, consider an organ pipe with a certain length L. When air is blown into the end of the pipe, the turbulence generates sound waves with many different wavelengths running back and forth between the ends of the pipe, overlapping and interfering with one another. Eventually all wavelengths except one will die out. The survivor's wavelength is 2L, that is, half its wavelength fits nicely in the pipe. Actually, harmonics of this wave, with wavelengths L, 2L/3, L/2, etc. also fit and also survive, and all form "standing waves." The pipe acts as a *resonant cavity,* accumulating sound energy only at specific wavelengths or frequencies.

2.3 The Cutoff Frequency

Lamb's original theory applied to an atmosphere in which all quantities are constant on horizontal planes, but vary vertically. The force of gravity causes the gas to settle in a predictable way, namely, the gas pressure falls off exponentially with altitude. This means that with each rise in altitude equal to a certain distance, the "scale height," the pressure decreases by a factor of $e = 2.73$. In Lamb's theory, the gas temperature was assumed constant everywhere and as a result the scale height was also constant.

The frequency F and wavelength L of a sound wave are related to the speed of sound, S, by a simple formula: $S = F \times L$. At any place in the atmosphere, the local temperature fixes the sound speed, so that the higher the frequency, the shorter the wavelength.

If a sound wave's frequency exceeds the cutoff, so that its wavelength is much shorter than the constant scale height, the atmosphere appears to be nearly uniform in pressure and the wave progresses freely. But if its frequency is below the cutoff, its wavelength is long compared to the scale height, and so the atmosphere acts as a series of reflecting

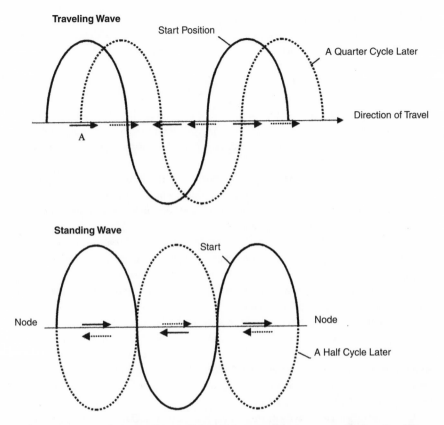

Traveling Wave

Start Position

A Quarter Cycle Later

Direction of Travel

A

Standing Wave

Start

Node

Node

A Half Cycle Later

FIG. 2.6   A standing sound wave (bottom) is trapped between two fixed nodes. All points between the nodes oscillate synchronously; that is, the phase advances by the same amount at all points in a given time. The oscillation is back and forth along the line between the nodes. The solid curve shows maximum amplitude of the oscillation at each point, the arrows show the direction of motion. A half cycle later (dotted line and dotted arrows), the directions have reversed, but the oscillations at all points are still in phase. In contrast, every point along the path of a traveling wave (top) has a different phase at any instant. Each point oscillates with the same period, but its phase lags behind that of its neighbor to the right. The net result is that the entire waveform shifts to the right.

walls. The incident wave will combine with its own reflections to form an attenuated standing wave, or *evanescent* wave.

2.4 *Fourier Analysis*

In connection with his studies of the flow of heat, Jean-Baptiste-Joseph Fourier, a French mathematician of the eighteenth century, showed that any arbitrary curve with definite endpoints could be represented as a sum of trigonometric functions (sines and cosines). The curve need not be oscillatory, but it nevertheless contains many different oscillations, with different strengths. His method and point of view had a huge impact on the subsequent development of mathematics.

2.5 Wave Types and Diagnostic Diagrams

In a gravitating atmosphere, like that of the Sun or Earth, several kinds of waves can exist. We have already mentioned *sound waves* in connection with Lamb's cutoff frequency, and they may be either propagating or nonpropagating. In addition, there are internal gravity waves and surface gravity waves.

Sound waves transmit changes in *pressure*. Snap your fingers and you launch a pressure wave in all directions that travels at the speed of sound, which depends on the air temperature. The air molecules oscillate back and forth a short distance (the wave amplitude), in the same direction as the wave propagates.

A second type of wave is the *internal gravity wave*, in which the restoring force is not pressure (as in sound) but *buoyancy*. When a bomb explodes, we hear a noise (sound waves) and also see the smoke boil upward because of the hot rising bubble of air the explosion produced. The bubble lifts the atmosphere above it and sends gravity waves propagating to the side and above. In such a wave, the air molecules oscillate *transverse* to the propagation direction, which is usually at an angle to the vertical.

Finally, there is the *surface gravity wave*, a transverse wave, which appears at the boundary between two fluids or gases of different densities. Ocean waves are a good example. Here gravity supplies the restoring force needed to maintain the oscillation. Because of the steep drop in gas density near the sun's "surface," such gravity waves are also present in the oscillation observations.

The temperature and gravity of the atmosphere exclude certain combinations of period and wavelength for each of these three types. In an isothermal atmosphere the separations are quite clear and can be displayed conveniently in a diagnostic diagram (see fig. 2.1). Sound waves with frequencies above a critical cutoff can propagate freely with no appreciable loss, but those with lower frequencies are evanescent. They are reflected by the vertical gradients of pressure, and so their amplitude decreases along their paths. The curved line at the top of the diagram separates these two classes of waves. Note that the horizontal wavelength also helps to determine which waves can propagate. Another cutoff or limiting frequency applies to internal gravity waves. They can only have frequencies *below* this "buoyancy" cutoff. Finally, the evanescent (nonpropagating) sound waves occupy the region between the two limiting curves in the diagnostic diagram.

2.6 Refraction of Sound Waves

Figure 2.7 shows a sound wave that is moving down into the Sun at an angle to the vertical. Each line represents a short "wave front" on which the phase is constant. Because the gas temperature and the sound speed increase inward, the lower end of each wave front will travel a little faster than the upper end. The result is that the wave front pivots, like a

Surface

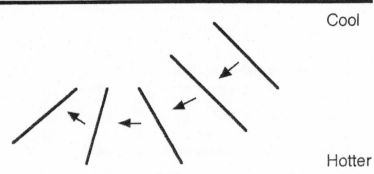

Cool

Hotter

FIG. 2.7   A plane wave moving down at an angle, into a region of increasing temperature (and therefore, sound speed), is bent back toward the surface. This is the phenomenon of refraction, which is common to all types of waves.

line of soldiers making a turn in a parade. The wave turns gradually until it moves horizontally at a "turning depth" and then back toward the surface. This is the phenomenon of
refraction, which occurs in all waves, including light waves and sound waves.

Note that the turning depth depends on two things: the particular variation of temperature with depth, and the wave's angle to the horizontal. The angle determines the horizontal and vertical components of the wavelength. All waves moving at an angle will be
refracted eventually, but only some waves can form a standing wave pattern.

A wave can be trapped as a standing wave *only* if half of its vertical wavelength
(the vertical component of its actual wavelength) fits an integral number of times
($n = 1, 2, 3, \ldots$) between the surface and the turning depth. The vertical wavelength
depends, in turn, on the wave's frequency and horizontal wavelength.

3 • A CLOSER LOOK AT SOLAR OSCILLATIONS

3.1 How to Unscramble the Modes on the Sun

Imagine that every minute for the past hour we have recorded a velocity map like the one
in figure 3.1. Remember that this map is a superposition of a large number of modes, all
oscillating at the same time. Now we want to determine which modes are present at the
surface, as well as the frequency and strength of each mode.

The first step is to correct the maps for a variety of extraneous effects, such as the slow
rotation of the Sun during the hour, the Earth's rotation and motion in its orbit, and so
on. Next, we pass the maps through spatial filters, one for each (L, M) mode. Some of the
filters are shown in figure 3.3. Imagine that the bright areas are perfectly transparent and
the dark areas are completely opaque. In practice, we multiply each of our sixty maps with
a particular filter and sum the products over the whole disk to get a measure S (L, M) of

the strength of that mode. Then we Fourier analyze the resulting time series of 60 S (L, M)s in order to determine the frequency associated with that mode.

3.2 How to Build a Toy Sun

A model of the Sun, in this context, is a table of numbers that lists the temperature, density, pressure, and other auxiliary quantities at each depth in the Sun. The model must reproduce the Sun's observed radius and luminosity at its present age, and if possible also predict the observed flux of solar neutrinos. The ultimate model that astronomers seek will also predict the observed oscillation frequencies of the Sun with high accuracy.

A "standard" model is based on the following physical principles and assumptions, each represented by an equation or mathematical constraint:

a. The Sun is a static, nonrotating sphere of gas. Its original chemical composition (mainly hydrogen and helium) was similar to that of comets.
b. The Sun has neither gained nor lost mass during its lifetime.
c. The Sun is in hydrostatic equilibrium. That means that the pressure at every depth balances the weight of the gas above it.
d. The Sun is in thermal equilibrium, at least outside its core. That means that energy leaves each volume at the same rate it enters. In the core, energy is released by the fusion of protons to form helium nuclei by one or more chains of thermonuclear reactions.
e. The Sun is in thermodynamic equilibrium. That means that, in every volume of gas, the atomic states and reaction rates are the same as though the volume were enclosed in perfectly reflecting walls at a uniform temperature. (The gas is so opaque to radiation that this assumption is reasonable.)
f. Energy is transported through most of the Sun's interior by radiation (X rays). However, in a spherical shell below the surface, approximately 0.3 R thick, convective motions of the gas carry a large fraction of the energy toward the surface.

To start with, a trial model of the Sun is adopted, with the present mass and an original composition. Most standard models differ principally in the amounts of helium and heavy metals assumed for the initial composition of the Sun. All the initial physical quantities as well as the nuclear rates at the core are specified, and then the model is "evolved" in a computer. That is, the temperatures and densities everywhere are allowed to adjust to the changing composition and energy production rates in the core. This evolution proceeds up to the present age of the Sun, 5 billion years. If the final model predicts the present solar luminosity, the model is acceptable. If not, the original model or the basic assumptions must be modified.

The present radius of the Sun can always be predicted correctly by adjusting one free parameter, the "mixing length" of hot bubbles in the convection zone.

3.3 Three Critical Frequencies

The Sun's internal structure is determined by four basic equations that express the principles mentioned in note 3.2. Three of these equations express the conservation of mass, momentum, and energy. (For example, mass is neither created nor destroyed.) The fourth equation (of "state") relates the pressure, temperature, and density of the gas.

In a perturbation analysis, every physical quantity (e.g., the pressure) at each depth is assumed to consist of two parts: the equilibrium value that a standard model requires, and a small fluctuating part. The three conservation equations require that certain relationships hold among these small perturbations. If in addition the perturbations are required to be oscillatory, then several simple relations emerge among a wave's frequency, wavelength, and three critical frequencies. These frequencies vary with depth in the sun because they depend on the local temperature and density.

A small mass of gas oscillates with the *buoyancy frequency*, N, if it is displaced from its equilibrium position in the Sun and if it doesn't exchange heat with its surroundings as it bobs up and down. Basically, the blob overshoots its equilibrium position as it rises or sinks. The buoyancy frequency depends on the local acceleration of gravity and the radial gradients of pressure and density. The buoyancy frequency is the *maximum* frequency a traveling internal gravity wave can have and is shown as the heavy line in the upper right of figure 3.7.

The *Lamb frequency* is best described by its reciprocal, the *Lamb period*, the time a wave takes to oscillate between the surface and a specified depth, if it always travels with the speed of sound appropriate to that depth. The Lamb frequency also depends on L, the "degree." It is shown by the small dashed lines in figure 3.7.

Finally the *Lamb cutoff frequency* is the *minimum* frequency for a traveling sound wave. It depends on the speed of sound and the pressure "scale height," the distance the pressure falls by a factor of e = 2.73. . . . The cutoff is shown as the heavy dashed curve in figure 3.7.

4 • THE SCRAMBLE FOR OBSERVATIONS, 1975–1985

4.1 The Deficit of Neutrinos

In 1964, Raymond Davis, a physicist at the Brookhaven National Laboratory, set up an experiment deep in an abandoned mine to measure the Sun's output of neutrinos. These ghostlike elementary particles are produced as by-products of the thermonuclear reactions in the Sun's core. They interact so weakly with matter that most pass through the Sun and Earth without ever suffering a collision. But, just by the law of averages, a few do get stopped.

Davis trapped solar neutrinos in a large tank of cleaning fluid where they formed a radioactive type of argon. Using exquisitely sensitive radiochemical techniques he could detect a few neutrino collisions a month. After several years of patient work, he was able to announce that his measured flux of neutrinos was less than half of that predicted by the best solar models. It took a decade to resolve this discrepancy and in the end the answer turned out to involve not only solar physics but elementary particle physics. We'll return to this subject in chapter 7.

4.2 The Resonance Scattering Cell

The cell contains a vapor of sodium or potassium. These metal atoms scatter, absorb, and emit light particularly well at specific wavelengths, the spectral lines. (The yellow light of

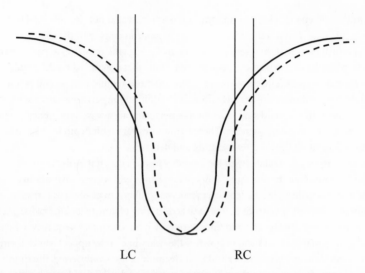

LC   RC

FIG. 4.6   The principle of the resonance cell. A sodium vapor, immersed in a magnetic field, absorbs light in a sodium spectral line only in the two bands (LC and RC) and
only if the incident light is circularly polarized. As the solar surface oscillates, the solar
spectral line shifts to the red (dashed line) and back to the blue ends of the spectrum. A
device in the instrument modulates the polarization of the incident light between left circular (LC) and right circular (RC), switching between the two bands. At any moment, the
difference of the intensity in the two bands yields the Doppler shift, which is proportional
to the velocity.

sodium vapor lamps on highways is emitted almost entirely within two lines, for example.) If the vapor is immersed in a magnetic field, the atoms will absorb light *only* at two
bands on each wing of the selected line of sodium (fig. 4.6). Moreover, in order to be absorbed, incident sunlight must have the correct circular polarization, clockwise for the
longer wavelength (RC), counterclockwise for the shorter wavelength (LC).

    In the instrument, a modulator polarizes the incident solar line of sodium alternately
clockwise and counterclockwise, at a high cadence. The difference in the amount of light
absorbed in each band is a measure of the Doppler shift or Doppler velocity. For example,
if the velocity averaged over the solar disk is zero with respect to the observer, the observer
will measure the *same* intensity of light in the blue and red wings of the spectral line. But
if the average velocity is *not* zero and away from the observer, the Doppler effect will shift
the line in wavelength toward the red, and the observer will measure a *lower* intensity for
clockwise polarization than for counterclockwise polarization. The difference in intensity
is a measure of the receding velocity, averaged over the solar disk.

    If the modulator runs at a steady cadence, the cell will record the Doppler shifts, aver-
aged over the solar disk, at many frequencies simultaneously. From this signal the spectrum of oscillations can be derived. Figure 4.2 is an example.

4.3 Cosmic Helium

According to current theory, all of the helium in the universe was originally created during the first few microseconds of the Big Bang. This event fixed the initial relative proportions of helium and hydrogen everywhere. Later this helium was incorporated into young stars, which have then proceeded to convert some of their hydrogen into helium. When some of these stars explode as novas, the helium is recycled. Therefore, a second or third generation of stars may have a slightly *larger* proportion of helium than during the Big Bang, but not a smaller proportion. For this reason a helium deficit in the Sun raises questions.

4.4 The Gravitational Deflection of Light

During 1974 and 1975, radio astronomers E. B. Fomalont and R. A. Sramek observed the occultation of three quasars by the Sun. The quasars lie close together in a straight line in the sky and the Sun passes among them. By observing at two high radio frequencies, they were able to eliminate the refraction of the radio waves from the quasars caused by the Sun's atmosphere. They obtained a deflection within one percent of Einstein's prediction of 1.76 arc-seconds. Observations with the Very Long Baseline Interferometer have reduced the uncertainty to 0.2%.

4.5 The Variation of the Solar "Constant"

ACRIM observations of the total irradiance of the Sun show a decrease of 0.1%, in step with the eleven-year sunspot cycle. The competing effects of bright "facular" regions and dark sunspots, both of which are more numerous at solar maximum, seem to explain this variation.

4.6 Kotov's Method

Kotov's magnetograph was originally intended to measure solar magnetic fields, but a slight modification allowed him to measure velocity oscillations instead. The instrument was basically a spectrograph, and he measured Doppler shifts of a spectral line to derive velocities in the usual way. In order to eliminate the slow wavelength drifts of the instrument, he used a differential method. He measured the average velocity in a strip that extended from pole to pole along the central diameter of the solar disk. Then he subtracted from that the average velocity over the rest of the disk, which was closer to zero, and retained those differences. The same wavelength drift would affect both measurements equally, but the drift would cancel when the subtraction was made. In this way he could obtain a long time series of oscillation data with a stable zero point.

5 • WHEELS WITHIN WHEELS: THE SUN'S INTERNAL ROTATION

5.1 Why Mode Frequencies Split

Figure 4.3 shows an actual example of frequency splitting. Why does this occur and what can we learn from it?

In figure 3.3 we saw a variety of spatial patterns, designated by the numbers L and M, in which the surface of the Sun can oscillate at a characteristic frequency, say F. Let's take a simple example, the striped one with L = 7 and M = 7. In the figure, white areas are rising at this moment and dark areas are falling. The colors will reverse in half a period.

Let's imagine an astronomer observing this striped pattern. She's selected a large rectangular area on the solar disk that includes many stripes. Now she's recording the Doppler velocity oscillations in each pixel in the rectangle. While she observes, the Sun is rotating, and therefore the pattern of stripes drifts from east to west within her fixed rectangle. The drift is a direct result of the ability of the solar gas to entrain sound waves.

Now imagine that we freeze the oscillations at some instant, so that the velocity in each stripe is constant, and that we also allow the Sun to continue to rotate. Although we have quenched the Sun's oscillations, the astronomer will nevertheless detect a periodic signal, because the stripes are moving through her fixed observing area. (The effect is similar to riding past a picket fence and seeing the scene behind it flicker.) The frequency she records is equal to the number of stripes, M, around the equator, divided by the rotation period, P (or F = M/P).

Next, suppose we freeze the Sun's rotation and also allow it to resume oscillating. The astronomers will then, obviously, record the oscillation frequency, F. With *both* rotation and oscillation working, she would record the *sum* of the two frequencies, or F + M/P.

As we see from figure 3.3, the rules of standing waves allow only certain values of M for a particular L. If L = 7, for example, the allowed values of M are $-7, -6, -5, -4, -3,$ $-2, -1, 0, 1, 2, 3, 4, 5, 6, 7$. Our observer will see a pattern corresponding to values of M from -3 to $+3$. The central peak corresponds to M = 0.

5.2 Inverting the Data to Derive Physical Quantities

Here is a very simplified sketch of the basic idea. Suppose we want to determine a quantity like the sound speed c(r) that varies only in one dimension, the radial direction r.

The observed period p of a sound wave is a kind of weighted average of the travel time $\Delta r/c(r)$ across each distance Δr between the surface, r = R and the turning radius of the mode $r = r_t$. The average can be written as a sum of products: $p = \Sigma[K(r) \times \Delta r/c(r)]$.

The weighting factor K determines the strength of the mode's oscillation at each radius r. Before we can proceed, we need to calculate K(r) using an independent (and reasonably accurate) model of the Sun's temperature and density.

Some modes penetrate deep into the Sun, others turn around at a shallow depth. If we add a number of K's wisely, their sum (K1 + K2 + K3 + . . .) will have a peak at some radius r* and will almost vanish everyplace else.

So if we add the frequencies F1 + F2 + F3 + . . . of those chosen K's, we find F1 + F2 + F3 + . . . = $\Sigma[(K1 + K2 + K3 . . .) \times c(r)]$. This is approximately equal to c(r*), the speed of sound at a single radius r* because the sum of K's is almost zero everywhere but at r*. We can repeat this process if we have enough modes of the right behavior, and so construct, step by step, a plot of the sound speed at each radius.

A similar process applies to the angular rotation speed, but it is more complicated because the speed varies in two dimensions, radius and latitude.

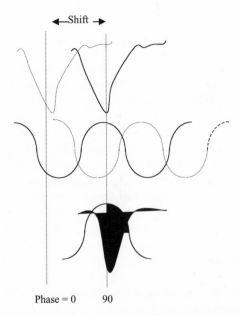

FIG. 5.9 Operation of the Fourier tachometer. The top panel shows a solar absorption line at its undisplaced (dotted line) and Doppler-shifted (solid line) wavelengths. The middle panel shows the sinusoidal pattern of optical transmission as it sweeps in wavelength. The bottom panel illustrates the product of the intensities in the spectrum and the transmission curve at the instant of maximum transmission.

5.3 Fourier Tachometer

The key element in the tachometer is a Michelson interferometer. The amount of light it transmits varies sinusoidally with the wavelength of light (fig. 5.9, middle). So when a spectrum is passed through the instrument, a pattern of light and dark stripes is created. By suitable means, this pattern can be swept smoothly in wavelength, so each wavelength of light is periodically illuminated. (The dotted and solid sinusoids in the middle panel show the pattern at two different moments.)

In practice, a short spectrum near a solar spectrum line is passed through the device (fig. 5.9, top). As the sinusoidal transmission pattern sweeps rapidly across the line, the amount of transmitted light is monitored. At the moment the amount reaches a maximum, the *phase* (from zero to 360 degrees) of the sinusoid is recorded. This phase, relative to the phase at a reference laser wavelength, is then a measure of the wavelength shift of the solar line. As the whole process is repeated periodically, a solid-state camera records the Doppler shift simultaneously for all points on the solar disk, as a function of time.

5.4 The Magneto-Optical Filter

The filter consists of two sodium vapor cells in series but uses transmitted, not scattered, solar light to measure velocity. An image of the Sun is presented to the device. A short piece of the solar spectrum of this image, containing the sodium absorption line at 569.4 nm, is first linearly polarized, then passed to the first cell. The sodium atoms in the magnetic field can absorb light only in two narrow bands (one on each side of the line center) because of the Zeeman effect. The light that is absorbed is circularly polarized as it passes through the cell. When the beam passes through the second cell, all the linearly polarized light outside the two bands is absorbed, leaving only the two bands. The net result is a very clean signal with a low background. Each band is turned on and off in turn. From the difference of light intensity in the two bands, one obtains the Doppler velocity (as in fig. 1.7). Because the device accepts a complete image, its output is the velocity at all points on the solar disk. The filter is very stable, hardly drifting in wavelength at all, which, as we will see, made it the perfect tool for low L studies.

7 • NEUTRINOS FROM THE SUN

7.1 The Proton-Proton Chain

The first step, which we saw earlier, is the collision of two protons with the formation of a deuteron, which consists of a proton and a neutron. Neutrons and protons have nearly the same masses, and one can change to the other by gaining or losing a unit positive electric charge. In this reaction, the positive charge of one proton is carried off by a new particle, a positron, which is a positively charged electron. In addition, a neutrino, a particle nearly without mass or charge, is created. We can write the reaction as

$$^1H + {}^1H \rightarrow {}^2D + e^+ + \nu_e.$$

The symbols correspond to ^1H for a proton (the nucleus of the hydrogen atom), ^2D for the deuteron (with a mass of two protons), e^+ for the positron, and ν_e for the neutrino. The three final particles carry away the released energy in the form of kinetic energy, which amounts to 1.4 million electron volts (MeV). The neutrinos have a maximum energy of 420 kilovolts.

Once a deuteron is formed it may combine with another proton to form a lightweight version ("isotope") of helium, ^3He, with a mass of three protons, as follows:

$$^2D + {}^1H \rightarrow {}^3He + \gamma.$$

The gamma (γ) in the above equation indicates that a gamma ray (a hard X ray) is released, with an energy of 5.5 MeV. From here on there are two paths to complete the chain. In the first, two ^3He combine to form the standard form of helium, ^4He, with a mass of four protons:

$$^3He + {}^3He \rightarrow {}^4He + {}^1H + {}^1H$$

Two protons are ejected and the three final particles carry away kinetic energy of 12.9 MeV.

In the second path, a beryllium nucleus ^{7}Be is formed. It picks up an electron and converts to a lithium nucleus ^{7}Li and then to ^{8}Be, as follows:

$$^3He + {}^4He \rightarrow {}^7Be + \gamma$$

$$^7Be + e^- \rightarrow {}^7Li + \nu_e$$

$$^7Li + {}^1H \rightarrow {}^8Be + \gamma + {}^4He + {}^4He.$$

Alternatively, the chain may end with the formation of a boron nucleus, ^{8}B:

$$^7Be + {}^1H \rightarrow {}^8B + \gamma$$

$$^8B \rightarrow e^+ + \nu_e + {}^4He + {}^4He.$$

The result of all these reactions is to fuse four protons into a helium nucleus, ^{4}He, with the release of two positrons, e^+, two neutrinos, ν_e, and up to 26.7 megavolts of energy in the form of gamma rays and the kinetic energy of created particles. The latter is redistributed by collisions very quickly among all neighboring particles and serves to maintain the local temperature.

7.2 The Physics of a Solar Model

The three basic equations governing a star in equilibrium, like the Sun, describe the conservation of mass, energy, and momentum. In addition, an "equation of state" that relates the gas temperature and density is needed. One assumes usually that no mass is lost or gained during the evolution of the Sun and ignores all macroscopic motions except those of convection. Spherical symmetry is assumed.

As input, one needs the mass and original chemical composition of the interstellar gas cloud in which the sun formed. Also, one needs the nuclear reaction rates for the proton-proton cycle as well as tables of the opacity of the gas to any wavelength, as a function of temperature and density. Convection is treated with a standard "mixing length" theory that contains a single adjustable parameter. At the center of the Sun, the gas temperature and density must remain finite, while at large distances from the center they must approach zero.

The equations are advanced in time steps, and all physical quantities are recalculated. At a simulated time equal to the present age of the Sun, the calculated radius and luminosity must be correct for the model to be valid.

7.3 The Solar Neutrino Unit

A SNU is defined as 10^{-36} neutrino captures per second, per target atom.

7.4 Calculating Oscillation Frequencies

The first step is to calculate an equilibrium model, as in note 7.2. Then we add a small time-dependent "perturbation" to each physical quantity that appears in the three conservation equations. For example, the pressure P becomes $P_0 + p(r, t)$, where $P_0(r)$ is the pressure in the equilibrium static model. The equations become simpler. That is, certain terms drop out because the perturbations are assumed to be small, p much less than P_0, for example. Then periodic solutions are sought, and these require specific relationships among the wave frequency, sound speed, and three characteristic frequencies. These are the Lamb cutoff frequency ω_c, the Lamb frequency Lc/r, and the buoyancy frequency N. Their depth variation determines the regions where a wave with frequency ω is trapped (see fig. 3.7). Two kinds of waves, with different restoring forces (pressure or gravity), have their own separate trapping regions.

8 • PICTURES IN SOUND

8.1 The Coriolis Force

Imagine that we fire an artillery shell due north from the equator. If the Earth were not spinning, the shell would land at a point (call it "a") due north of the gun. But because the Earth spins, as seen by a "stationary" observer on a distant star, the gun and the shell (as it leaves the gun) move eastward at the linear speed of rotation at the equator, namely 1600 km per hour. All points at higher northern latitudes (including the point "a") have lower linear speeds. The pole, for example, doesn't rotate at all. Therefore, by the time the shell lands, it has traveled farther east than point "a." An observer at point "a" would say that some force (like a wind) has deflected the shell to the east, whereas in fact no force but gravity (directed downward, not eastward) was acting on the shell. This apparent force, due only to rotation, is called the *Coriolis force* after the French mathematician Gustave Gaspard Coriolis (1792–1843). As seen by observers on the rotating Earth, it deflects moving objects to the right in the Northern Hemisphere and to the left in the Southern Hemisphere. Thus, it causes the wind to rotate counterclockwise as it approaches a low-pressure zone in the Northern Hemisphere and clockwise in the Southern Hemisphere.

8.2 Magneto-Sonic Waves

Three basic kinds of waves can propagate in a uniform magnetic field. Ordinary *sound waves*, in which gas pressure is the restoring force, propagate *parallel to* the direction of the magnetic field at the speed of sound, c. Sound waves also oscillate in the direction of the field. *Alfvén waves* travel along the field lines at the Alfvén speed, C. Their restoring force is the tension in the field lines, and their oscillations are *transverse* to the direction of propagation, like waves on a string. Finally, *magneto-sonic waves* travel *across* the field lines at a speed that is a function of C and c, namely $V = (C^2 + c^2)^{1/2}$. Their restoring force is a combination of magnetic and gas pressures.

9 • ROTATION, CONVECTION, AND HOW THE TWAIN SHALL MEET

9.1 Evolution

Charles Darwin's theory of evolution of biological species has two essential factors: variation and selection. A species' genes combine in different ways to produce individuals with slightly different abilities. The stresses of the environment select those individuals capable of survival. Only these individuals live to mate and reproduce their favorable qualities. In this way the species slowly changes to adapt to changes in the environment. Eventually, the genetic changes are sufficient to create a new species.

9.2 The Rayleigh-Taylor Instability

Imagine a quiet tidal pool of seawater. In the hot Sun, water in the top layer of the pool evaporates and so the layer grows saltier. The saltier the water in this layer, the denser it becomes. All is well until a bug lands on the pool. At this instant the water surface under the bug has a dimple. The heavier salty water will tend to slide into the dimple, making room for the lighter water underneath to rise buoyantly around the dimple. This is the instability, and it will grow rapidly. The dimple will deepen into a long finger of downflowing saltier water.

10 • THE SOLAR DYNAMO

10.1 Random Walk of Magnetic Flux

The classic example of a random walk is the progress of a drunk person who takes each step in an arbitrary direction. His distance from his starting point increases slowly, as the square root of the number of steps. Supergranules appear randomly over the surface of the Sun. Each new supergranule can shift the magnetic flux in its neighborhood by half its diameter (a "step"), on average. In this way the flux is distributed over the surface, away from high concentrations to more empty areas.

10.2 A Surface Dynamo?

If you look carefully at figure 9.1, you'll notice that the speed of rotation also varies with depth in a thin layer near the surface of the Sun. The speed increases inward at 60-degree latitude and decreases inward at zero and 30 degrees. In principle, this layer could also serve as a site for an alpha-omega dynamo. In fact, Peter Wilson, an Australian theorist, has proposed that the small-scale, random magnetic fields that are observed all over the Sun could be generated in this layer. One question with this idea is whether a surface dynamo could generate the field strengths (hundreds of gauss) one sees at the surface.

10.3 Buoyancy and the Temperature Gradients in the Sun

In the radiative zone the negative temperature gradient is determined by the flow of energy (X rays) to the surface. The gradient in the zone is too flat to permit a blob of gas to rise buoyantly. If the blob rises quickly, it expands and cools "adiabatically," that is, without exchanging energy with its surroundings. To continue to rise, the blob's temperature must always be higher than its surroundings. But if, as in the radiative zone, the temperature gradient is flatter than the adiabatic gradient that the blob follows, the blob arrives cooler and therefore denser than its surroundings. It will therefore sink. So, blobs are not buoyant, or, in other words, convection is suppressed in the radiative zone. Magnetic fields are "frozen" in the gas and behave similarly.

10.4 Alpha and Omega Effects

As a convection cell rises, it creates a hairpin-shaped kink in the toroidal field, and forces it to rise, too. The Coriolis force tends to turn the cell and the kink in a clockwise direction in the Northern Hemisphere and in the counterclockwise direction in the Southern—the so-called alpha effect. The omega effect is the stretching and overlapping of field lines in a shearing flow of gas. As the field lines overlap, the field strength increases. The omega effect thus amplifies the field strength.

10.5 Explanation of the Charbonneau-MacGregor Model

Each panel of figure 10.9 shows the fields at a different phase of the cycle. In each panel, the quadrant on the left displays the toroidal parts of the field (which extend in longitude around the Sun), and the one on the right shows the poloidal parts of the field (which extend along meridians). In effect, the ropes are twisted, with some poloidal field lines poking through the surface. In the boxes below the quadrants we see an expanded view of the tachocline, with contours of a toroidal field on the left and a poloidal field on the right. Clockwise, poloidal field lines and positive toroidal fields are plotted in solid lines, and the reverse in dotted lines.

First, look at the left quadrants, where you will see two toroidal ropes of opposite magnetic polarity, moving down toward the equator. At first, the positive rope is stronger, but as it fades, the negative one takes over. Similarly, in the right quadrants, positive poloidal fields emerge near the poles and drift toward the equator. As they weaken, negative poloidal fields emerge.

The actual field is the sum of the fields in the two quadrants. In effect, the ropes are twisted. They have strong toroidal fields and weaker poloidal fields. When a rope rises buoyantly through the surface, it appears as a belt of twisted field, extending in longitude, which supposedly represents a sunspot belt.

In the left box of panel A, we see a stack of ropes of alternating polarities just below the interface at 0.7 radii. These ropes were generated in previous cycles and are waiting their turn to rise through the interface. On the right box of panel A, we see the poloidal field associated with the uppermost rope. This poloidal field was generated by the alpha effect in the convection zone and is diffusing downward across the interface. The strong shear below the interface converts the poloidal field into a toroidal field.

11 • AD ASTRA PER ASPERA — "TO THE STARS THROUGH ENDEAVOR"

11.1 Cepheid Pulsation

A Cepheid pulsates radially, expanding and contracting like a balloon. Observations show that a star reaches maximum brightness before it reaches maximum size, contrary to what one might expect. The reason is that the temperature rise more than compensates for the lesser surface area. We now understand that the pulsations are driven by the storage of radiant energy in the ionization energy of helium ions during the compression stage, and the release of that energy in the expansion stage. The pulsation is limited to the stellar envelope. The flow of energy from the core to the envelope is continuous, and is only interrupted periodically in the envelope.

11.2 The Hertzsprung-Russell Diagram

Einar Hertzsprung was a Danish astronomer who is credited with being the first to sort stars according to their color and brightness. Henry Norris Russell, a professor at Princeton University and a towering figure in twentieth-century astronomy, discovered the value of this famous diagram independently.

Most stars are located on a so-called main sequence, a sloping band in the diagram, in which the fainter stars are both redder and less massive. The most luminous stars can have any color from blue to red and are sprinkled across the top of the diagram. As a star on the main sequence matures, it can become a red giant. It then migrates in the diagram upward to greater brightness and to the right to redder color. Some stars then become blue and ultimately fade into white dwarfs.

11.3 Star Names

Only the brightest or most unusual stars are given proper names. The rest are designated in a number of ways, by catalog number or often by their location in a constellation. The brightest star is labeled "alpha" and the next brightest "beta" and so on. For example, Betelgeuse, the brightest star in the constellation Orion, is also called Alpha Orionis, Latin for "Alpha of Orion."

11.4 Degenerate Electron Gas

In an ionized gas, only two electrons, with opposite spins, can occupy a single atomic energy state. That is a statement of Wolfgang Pauli's exclusion principle, and it applies everywhere with no exceptions. In a normal star, there are plenty of available energy states that electrons can occupy. In a white dwarf, though, the gas is compressed so much by gravity that electrons fill all available energy states. At that point the star becomes degenerate. It cannot contract any further because there are no allowed locations that electrons can occupy.

11.5 Hipparchos Satellite

This satellite was designed to measure the positions and transverse motions of 100,000 stars in the Sun's neighborhood. The satellite used a familiar method, the measurement of "parallax," but made such measurements with unprecedented accuracy, a thousandth of an arc-second.

As the Earth revolves around the Sun, the apparent position of a star shifts, since it is viewed from different points along the Earth's orbit. The maximum angular shift, which is seen at opposite sides of the orbit, is the star's parallax. The parallax is smaller the farther away the star is, and from simple geometry its distance can be determined from its parallax. The Hipparchos satellite was launched in 1993 and finished its task in 1996. A comprehensive catalog of parallaxes was published in 1997 by the European Space Administration.

11.6 Distances of Star Clusters

A star cluster or association is a group that was born at about the same time from a single large interstellar gas cloud. Once the distance of one cluster has been determined, by any means, then finding the distances to similar clusters becomes simple. In effect, a Hertzsprung-Russell diagram is constructed for the new cluster and then overlaid on the diagram for the calibrated cluster. The vertical shift, along the brightness axis, is then a measure of the relative distance of the new cluster. This method averages out the random variations among stars of similar type and thus is more accurate than a distance measurement for a single star.

12 • SOME LATE NEWS

12.1 Why the Solar Constant Varies

Sunspots are darker than the average surface of the Sun because their strong magnetic field inhibits the flow of heat to the surface by convection cells. The heat that would reach the surface is blocked and is stored in the convection zone for months and released later over a broad area; that effect decreases the solar constant. In active regions, areas brighter than average (faculae) appear and add to the Sun's emission; that effect increases the solar constant. The two effects nearly balance but not quite, and so their sum varies as the numbers of spots and active regions vary through the solar cycle.

12.2 The Widths of Oscillation Peaks

Fourier theory tells us that only an oscillation that persists for an infinite length of time has an absolutely precise frequency (F). In contrast, a sinusoidal oscillation that has a definite beginning and end (that is, a lifetime, T), contains a small range of frequencies (ΔF) near its natural period, F. The frequency range is effectively the reciprocal of the lifetime $\Delta F = 1/T$.

GLOSSARY

acoustic	Relating to sound.
adiabatic change	A change in the state of a system in which no energy is exchanged.
Alfvén wave	A propagating transverse or torsional wave in a magnetic field.
amplitude	The maximum displacement of the medium during a cycle of an oscillation.
angular size	The width, in degrees or arc-seconds, of a distant object. The Sun, for example, is half a degree in width.
arc-second	An angle $1/3600$ of a degree, or 4.848×10^{-6} radians. An arc-second at the distance of the Sun subtends 725 km.
azimuthal order	In oscillations on a sphere, the number of nodes on the equator.
bipole	A pair of regions of opposite magnetic or electrical polarity.
blackbody	An object that absorbs all the radiation incident upon it. Blackbodies emit a characteristic spectrum regardless of their actual composition.
chromosphere	The layer in the atmosphere of the Sun above the visible surface.
convection zone	The layer in the interior of the Sun below the visible surface in which energy is transported partly as the internal heat of the gas and partly as radiation.
corona	The hot outer solar atmosphere, which extends many radii from the surface.
cosmic ray	A generic term for high-energy charged particles (e.g., protons) that impinge on the Earth from outer space.
degree (L)	In oscillations on a sphere, the number of nodes on a meridian.
diagnostic diagram	A graph of the power in an oscillation, as function of frequency and horizontal wavelength (or degree). See figure 3.5.
differential rotation	In the Sun, the variation of the angular speed of rotation (in, say, radians/sec) as a function of latitude and depth. It can also be expressed as a frequency of rotation (in microhertz).
diffraction	The tendency of a wave to bend around obstacles, or to spread out in passing through an aperture.

Doppler effect   The shifting of frequencies of light or sound that an observer receives from a moving source. The shift is to lower frequencies (or to the red end of the spectrum) with receding source. (Hence, the term "redshift".)

duty cycle   In a cycle (like night and day), the fraction of time during which useful observations or operations can be made.

dynamo   A mechanism for generating electrical current and magnetic fields.

effective temperature   The temperature of a blackbody at which the surface of a star would radiate its luminosity.

electron volt (eV)   The amount of energy an electron gains by passing through a voltage drop of one volt. Equivalent to $1.6 \times (10^{-19})$ watt-seconds. MeV = million eV.

evanescent   Transient, of short duration.

flare   On the Sun, a violent rapid release of radiation and charged particles, usually caused by the annihilation of a magnetic field.

Fraunhofer lines   Wavelengths in the spectrum of sunlight at which the brightness is reduced by absorption and scattering by particular atoms in the solar atmosphere.

frequency   The number of times per second (or some other unit of time) that some event recurs.

gauss (G)   A unit of magnetic field strength. The magnetic field at the Earth's magnetic poles is about half a gauss.

granule   A convection cell at the Sun's surface, with a typical size of 1500 km and lifetime of ten minutes.

gravity wave   A transverse wave driven by buoyancy and gravity. Internal gravity waves propagate in nearly homogeneous media, surface waves at a sharp discontinuity in the medium.

hertz (Hz)   A unit of frequency, equal to one cycle per second.

hydrodynamics   The science of the motions of fluids, including liquids and gases.

irradiance   The amount of solar energy received outside the Earth's atmosphere per square meter and per second.

lepton   One of a family of light elementary particles that includes the electron, tauon, muon, and three types of neutrino.

limb   In astronomy, the edge of the visible disk of the Sun, the Moon, or a planet.

luminosity   The rate at which the total energy of a star is emitted.

magnetic field   A region in space in which a moving electrical charge experiences a lateral force, which is proportional to its speed and charge, and to the "strength" (in gauss) of the field.

magnetic flux   In an area threaded by the magnetic field, the flux is the product of magnetic field strength and area. Loosely speaking, a measure of the quantity of magnetism. The unit is an oersted.

magnetograph   A device to measure and record magnetic fields. Most magnetographs measure magnetic flux rather than intensity or field strength.

meson	An elementary particle, with a mass about two hundred times that of the electron. The group includes the pion, kaon, and psi particles. They exist with positive, negative, or zero electric charge. The charged pions decay into muons and neutrinos.
microhertz	A frequency of one-millionth of a hertz, or one cycle in a million seconds.
millihertz	A frequency of one-thousandth of a hertz, or one cycle in a thousand seconds.
mode	A pattern of vibration.
model	A mathematical representation of a physical object or process. A "model" of the interior of the Sun, for example, would consist of a set of tables that specify such quantities as temperature and density as a function of radial distance.
muon	An unstable elementary particle with a mass 207 times that of the electron. The muon was originally thought to be a meson, but is now recognized as a lepton.
nanohertz	A unit of frequency, equal to a billionth of a hertz.
nanometer	A unit of length equal to a billionth of a meter.
neutrino	An elementary particle. It has no electric charge or magnetic moment. Three types exist, and at least one has a mass of about $1/60,000$ of an electron's mass.
node	A point in an oscillating medium at which the amplitude is zero. In a traveling wave the nodes propagate at a characteristic speed; in a standing wave they are fixed in space.
period	The interval of time between repetitions of a cycle.
phase	The state of an oscillating system at some point in its cycle. For example, the Moon has phases, passing from full to last quarter to new moon and so on.
photometer	A device for measuring light. Photodiodes, photomultipliers, and charge-coupled devices (CCDs) are examples.
plasma	A gas that consists of electrically charged particles. Positive and negative charges are present in equal numbers, so that the gas as a whole is electrically neutral.
p-mode	An oscillation in which pressure (p) is the restoring force.
poloidal field	A field whose field lines connect two separated magnetic poles. The field lines resemble the familiar pattern of iron filings near a bar magnet (a "dipole"). In the Sun the term usually refers to north-south fields.
radial order	The number of nodes of an oscillation that lie along a radius of the Sun.
refraction	The bending of a wave of light or sound, caused by a variation of the speed of propagation in the medium. A prism of glass refracts light waves, and a gradient of temperature in the Sun refracts sound waves.
resonance	The response of a vibrating system to a force that has the same frequency as the natural frequency of the system. The amplitude of the vibration increases exponentially at resonance. For example, pushing a child's swing in time with the swinging produces a wild ride.

resonant cavity   A closed space whose dimensions determine the wavelengths of a resonant oscillation.

shock wave   A large-amplitude propagating wave, in which energy is dissipating. A sonic boom is a familiar example.

solar cycle   The sun's cycle of magnetic activity. The number of sunspots varies with a period of about eleven years, but a complete cycle, in which the magnetic polarities of the Sun's poles reverse, lasts about twenty-two years.

sound wave   A propagating periodic disturbance of pressure in a gas or plasma. A point in the medium is displaced back and forth, along the direction of propagation, as the wave passes by.

spectral line   A narrow band of wavelengths at which a particular atom or ion emits and absorbs light strongly.

spectrograph   An instrument that spreads light into its constituent colors or spectrum. The most commonly used spectrographs use diffraction gratings to disperse the light.

spectroheliograph   An instrument that generates an image of the Sun in a narrow band of wavelengths.

spectrum   The full range of wavelengths or frequency emitted by a source, displayed in a linear order.

standing wave   A wave in which the nodes remain fixed in space during the oscillation.

supergranule   A solar convection cell with a typical size of 30,000 km and a turnover time of about a day.

tachocline   A thin layer between the convection and radiative zones, in which the angular speed of rotation varies both along a radius and in latitude.

tauon   An elementary charged particle, a member of the lepton group.

toroidal field   A magnetic field with the shape of a doughnut. In the Sun, toroidal fields extend in the east-west direction, that is, in the direction of longitude.

traveling wave   A wave that propagates in space while maintaining its form, like an ocean wave.

umbra   The darkest part at the center of a visible sunspot. The penumbra is the somewhat lighter ring region around the umbra.

wavelength   The distance between successive maxima (or successive minima) in a wave. The wavelength of visible light is about 500 nanometers, and that of audible sound about half a meter.

Zeeman effect   The splitting of a spectral line into a number of close-spaced components, caused by the presence of an external magnetic field around the emitting atom.

INDEX

CREDITS

Fig. 1.1: Courtesy of National Solar Observatory

Fig. 1.2: Courtesy of L. Delbouille, University of Liège

Figs. 1.3–1.4: J. W. Evans, *Astrophysical Journal* 136 (1962): 815, figs. 2 and 3. Reproduced by permission of the American Astronomical Society.

Fig. 1.6: Courtesy of National Solar Observatory

Fig. 2.1: Courtesy of E. Frazier and *Zeitschrift für Astrophysik*

Fig. 2.2: SOHO/MDI (ESA and NASA)

Fig. 2.3: R. Ulrich, *Astrophysical Journal* 162 (1970): 993, fig. 2. Reproduced by permission of the American Astronomical Society.

Fig. 2.4: F.-L. Deubner, *Astronomy and Astrophysics* 44 (1975): 371, fig. 3

Fig. 2.5: Courtesy of E. Rhodes

Fig. 3.1: SOHO/MDI (ESA and NASA)

Fig. 3.3: F. Hill, F.-L. Deubner, and G. Isaak, "Oscillation Observations," fig. 2 in *Solar Interior and Atmosphere*, ed. A. N. Cox, W. C. Livingston, and M. S. Matthews. © 1991 The Board of Regents. Reprinted by permission of the University of Arizona Press.

Fig. 3.4: SOHO/MDI (ESA and NASA)

Fig. 3.5: SOHO/MDI (ESA and NASA)

Fig. 3.6: J. Bahcall and R. Ulrich, *Reviews of Modern Physics* 60 (1988): 323, fig. 6

Fig. 3.7: J. Christensen-Dalsgaard and G. Berthomieu, "Theory of Solar Oscillations," fig. 2 in *Solar Interior and Atmosphere*, ed. A. N. Cox, W. C. Livingston, and M. S. Matthews. © 1991 The Board of Regents. Reprinted by permission of the University of Arizona Press.

Fig. 4.1: Courtesy of E. Rhodes

Fig. 4.2: Courtesy of J. Brookes and the Birmingham oscillation team

Fig. 4.4: Courtesy of G. Grec and E. Fossat

Fig. 4.5: Courtesy of M. Woodard and H. Hudson

Fig. 5.1: K. Libbrecht and C. Morrow, "The Solar Rotation," fig. 1 in *Solar Interior and Atmosphere*, ed. A. N. Cox, W. C. Livingston, and M. S. Matthews. © 1991 The Board of Regents. Reprinted by permission of the University of Arizona Press.

Fig. 5.3: K. Libbrecht and C. Morrow, "The Solar Rotation," fig. 8 in *Solar Interior and Atmosphere*, ed. A. N. Cox, W. C. Livingston, and M. S. Matthews. © 1991 The Board of Regents. Reprinted by permission of the University of Arizona Press.

Figs. 5.4–5.5: Courtesy of T. Duvall and J. Harvey

Figs. 5.6–5.7: K. Libbrecht and C. Morrow, "The Solar Rotation," figs. 3 and 7 in *Solar Interior and Atmosphere*, ed. A. N. Cox, W. C. Livingston, and M. S. Matthews. © 1991 The Board of Regents. Reprinted by permission of the University of Arizona Press.

Fig. 5.8: P. Gilman et al., *Astrophysical Journal* 338 (1989): 528, fig. 1. Reproduced by permission of the American Astronomical Society.

Fig. 6.1: Courtesy of the BISON team

Fig. 6.2: F. Hill, F.-L. Deubner, and G. Isaak, "Oscillation Observations," fig. 3 in *Solar Interior and Atmosphere*, ed. A. N. Cox, W. C. Livingston, and M. S. Matthews. © 1991 The Board of Regents. Reprinted by permission of the University of Arizona Press.

Figs. 6.3–6.6: Courtesy of the GONG team

Fig. 6.7: SOHO (ESA and NASA)

Fig. 7.1: Courtesy of J. Bahcall

Fig. 7.2: R. Davis and A. N. Cox, "Solar Neutrino Experiments," fig. 3 in *Solar Interior and Atmosphere*, ed. A. N. Cox, W. C. Livingston, and M. S. Matthews. © 1991 The Board of Regents. Reprinted by permission of the University of Arizona Press.

Fig. 7.3: Courtesy of J. Bahcall and M. Pinsonneault.

Fig. 7.4: J. Bahcall, *Physics Today* 49 (1996): 30, fig. 3. Reproduced with permission of the American Institute of Physics.

Fig. 7.5: D. Gough et al., *Science* 272 (1996): 1298, fig. 5. Reprinted with permission of the American Association for the Advancement of Science.

Fig. 7.6: A. Kosevichev et al., *Solar Physics* 170 (1997): 73, fig. 10. Reproduced with kind permission of Kluwer Academic Publishers.

Fig. 7.7: Courtesy of B. Fleck, *ESA Bulletin* 102

Fig. 7.8: S. Basu et al., *Monthly Notices, Royal Astronomical Society* 292 (1997): 243. Reproduced with permission of Blackwell Publishers.

Fig. 7.9: Courtesy of J. Bahcall and M. Pinsonneault

Figs. 8.1–8.2: F. Hill, *Astrophysical Journal* 333 (1988): 996, figs. 10 and 3. Reproduced by permission of the American Astronomical Society.

Fig. 8.3: J. Patrón et al., *Astrophysical Journal* 455 (1995): 746, fig. 7. Reproduced by permission of the American Astronomical Society.

Fig. 8.5: T. Duvall et al., *Nature* 362 (1993): 430, fig. 1. Copyright permission by *Nature*.

Fig. 8.6: T. Duvall et al., *Nature* 379 (1996): 235, fig. 1. Copyright permission by *Nature*.

Figs. 8.7–8.9: A. Kosevichev et al., *Solar Physics* 192 (2000): 159, figs. 2, 5, 7. Reproduced with kind permission of Kluwer Academic Publishers.

Fig. 8.11: C. Lindsey and D. Braun, *Astrophysical Journal* 485 (1997): 895, fig. 2. Reproduced by permission of the American Astronomical Society.

Fig. 8.12: C. Lindsey and D. Braun, *Nature* 287 (2000): 1799, fig. 1. Copyright permission by *Nature*.

Fig. 8.13: Courtesy of D. Braun

Fig. 9.1: Courtesy of J. Schou and J. Christensen-Dalsgaard, SOHO (ESA and NASA)

Fig. 9.2: W. Chaplin et al., *Monthly Notices, Royal Astronomical Society* 283 (1996): L31. Reproduced with permission of Blackwell Publishers.

Fig. 9.3: S. Tomczyk, *Bulletin of the Astronomical Society of India* 24 (1996): 245

Fig. 9.4: P. Charbonneau, *Astrophysical Journal* 496 (1998): 1015, fig. 7. Reproduced by permission of the American Astronomical Society.

Fig. 9.5: Courtesy of T. Corbard and the GOLF team, Structure and Dynamics of the Interior of the Sun and Sun-like Stars, SOHO 6 / GONG 98 Workshop, June 1–4, 1988, Boston, Mass., ESA SP-418

Fig. 9.6: W. Chaplin et al., *Monthly Notices, Royal Astronomical Society* 308 (1999): 405, fig. 3. Reproduced with permission of Blackwell Publishers.

Fig. 9.7: M. Miesch, *Solar Physics* 192 (2000): 59. Reproduced with kind permission of Kluwer Academic Publishers.

Fig. 9.8: R. Stein and Å. Nordlund, p. 391, fig. 5, in *Mechanisms of Chromospheric and Coronal Heating*, ed. P. Ulmschneider, E. R. Priest, and R. Rosner (1991). Reproduced with permission of Springer-Verlag.

Figs. 9.9–9.10: N. Brummell et al., *Astrophysical Journal* 473 (1996): 494, figs. 2 and 3. Reproduced by permission of the American Astronomical Society.

Fig. 9.11: Courtesy of M. Miesch

Fig. 9.12: Courtesy of J. R. Elliott, Structure and Dynamics of the Interior of the Sun and Sun-like Stars, SOHO 6 / GONG 98 Workshop, June 1–4, 1988, Boston, Mass., ESA SP-418

Fig. 9.13: Courtesy of P. R Woodward, University of Minnesota

Fig. 10.1: Courtesy of the TRACE consortium. TRACE is a collaboration between NASA and ESA.

Fig. 10.2: Courtesy of Big Bear Solar Observatory

Fig. 10.3: Courtesy of D. Hathaway

Fig. 10.5: D. Rabin, "The Solar Activity Cycle," fig. 19 in *Solar Interior and Atmosphere*, ed. A. N. Cox, W. C. Livingston, and M. S. Matthews. © 1991 The Board of Regents. Reprinted by permission of the University of Arizona Press.

Fig. 10.6: Courtesy of M. Stix

Fig. 10.7: P. Gilman et al., *Astrophysical Journal* 338 (1989): 528, fig. 1. Reproduced by permission of the American Astronomical Society.

Figs. 10.9–10.10: P. Charbonneau and K. MacGregor, *Astrophysical Journal* 486 (1997): 502, figs. 7 and 8. Reproduced by permission of the American Astronomical Society.

Fig. 11.1: Courtesy of A. Sandage

Fig. 11.2: Courtesy of J. Christensen-Dalsgaard

Fig. 11.3: S. Kawaler and M. Dahlstrom, "White Dwarfs," *American Scientist* (Nov.-Dec. 2000), figs. 1 and 2. Courtesy of Tom Dunne / *American Scientist*. Reprinted with permission of *American Scientist*, magazine of Sigma Xi, The Scientific Research Society.

Figs. 11.5–11.6: D. E. Winget, *Astrophysical Journal* 378 (1971): 326, figs. 2 and 4. Reproduced by permission of the American Astronomical Society.

Fig. 11.7: T. Brown et al., *Astrophysical Journal* 368 (1991): 599, fig. 3. Reproduced by permission of the American Astronomical Society.

Fig. 12.1: A. Kosevichev and V. Zharkova, SOHO/MDI (ESA and NASA)

Fig. 12.2: C. Froehlich, SOHO/VIRGO (ESA and NASA)

Fig. 12.3: Courtesy of T. Corbard

Fig. 12.4: R. Howard and B. LaBonte, *Astrophysical Journal* 239 (1980): L33, fig. 1. Reproduced by permission of the American Astronomical Society.

Fig. 12.5: J. Schou et al., *Astrophysical Journal* 505 (1998): 390, fig. 5. Reproduced by permission of the American Astronomical Society.

Figs. 12.6–12.7: R. Howe et al., *Science* 287 (2000): 2457, figs. 7 and 2. Reprinted by permission of the American Association for the Advancement of Science.

Figs. 12.8–12.10: M. Dikpati and P. Gilman, *Astrophysical Journal* 559 (2001): 428. Reproduced by permission of the American Astronomical Society.